PRIMARY PHYSICS

THE PRINCIPLES BEHIND ROMAN MACHINES

AF779834

Marti Ellen

Illustrated by Andrew Davies

foreword by Professor Richard Collins

I retired at the end of 2000 after a 40 year career as a scientist in industry, government and academia. If I had to do it all again, I wouldn't change a thing. For me, science was, and always will be, enthralling, fulfilling and great fun. It is very satisfying to understand, at the most fundamental level, why something happens. Also, there is nothing like the feeling when, as happened to me just a few times in my research career, you realise that you know something that nobody else in the world does.

I was extremely fortunate to have several passionate and dedicated teachers who excited me about science. The *Primary Physics* series of books will do the same thing for today's young people.

There is a wonderful quote, often attributed to Confucius, about teaching and learning:

I hear and I forget.
I see and I remember.
I do and I understand.

The *Primary Physics* books are all about doing and understanding. They deal with some of the very basic concepts in mechanics – measurement, mass, motion, force and energy. Through cleverly designed activities, teachers and students are led through the process of doing, observing, describing, explaining and, ultimately, understanding. As is appropriate for young children, the approach is qualitative (which is bigger, or faster?) rather that quantitative – there are not many equations here! However, the concepts developed are rigorous and correct. Andrew Davies' charming and friendly drawings add clarity to the material presented and help to make the whole experience fun.

The simplicity of the words and drawings in the books belie the amount of work needed to create them. Time and again over the several years that it took to write them, Marti came to me with ideas and drafts that I encouraged her to revise until they were rigorously correct at the level of presentation. Every time that I raised concerns about the accuracy of an explanation or the appropriateness of an experimental design, she went away and worked it over until we were both happy. The *Primary Physics* books are a tribute to Marti's persistence, dedication and passion.

When I tell someone that I had a career in Physics, one of the most common responses I get is along the lines: "I was never any good at that – I couldn't understand what it was about." I believe that mostly this occurred because the teacher did not understand the subject matter well enough to impart it effectively. Certainly, setting out to teach science with little or no background in the field can appear quite daunting. However, *Primary Physics* can enable teachers with little or no experience in Physics to do just that. In addition to leading students through experiments and projects, all of the books contain Teacher's Notes that put the activities into a broader context. These Notes also provide background material to help answer the inevitable questions that young people are so good at asking. And while on this subject, may I offer a word of advice: don't feel apprehensive about the possibility (actually, the certainty!) that a student will ask you a question that you cannot answer. This happened to me regularly throughout my entire teaching career, and it always exhilarated me. Finding yourself in this situation enables you to praise the student for his/her perception. You can also use it to stimulate productive class discussion. Perhaps most importantly, it increases your own understanding. And if an answer doesn't emerge out of all this, you can promise to have it at the next session!

The world would be a very odd place indeed if everybody decided to become a scientist. Introducing children to science at an early age is not about that, however. Scientists look at the world in a particular way: they observe, test, analyse, discuss, explain and draw conclusions. Thinking about things in this way is also important for people in a host of other professions – indeed, for everyone. Our society is stronger when its members can make informed assessments of what they hear, see and read and act in an appropriate way. Being exposed to science at an early age will help to develop this powerful way of thinking. And some of those who are fortunate enough to learn about science when very young may well be stimulated to have exciting and rewarding careers in this wonderful field. As Henry Adams perceptively said:

A teacher affects Eternity: he (she) can never tell where his (her) influence stops.

Emeritus Professor Richard (Dick) Collins
Professor of Applied Physics,
University of Sydney 1980 – 2000
Head, School of Physics,
University of Sydney 1997 – 2000

introduction

Contents

First published March 2012
by INT Press Pty Ltd

This edition published July 2012
by Sunshine Educational Pty Ltd

Primary physics, the principles
behind Roman machines

Marti Ellen

ISBN 978-0-9586701-5-9

© 2017 Sunshine
Educational Pty Ltd
www.primaryphysics.com

This book is copyright. Apart from any fair dealing for the purposes of private study, research, criticism, or review, as permitted under the copyright act, no part may be reproduced by any process without written permission.

Educational institutions copying any part of this book for educational purposes under the Act must be covered by a Copyright Agency licence for educational institutions and must have given a remuneration notice to Copyright Agency. Licence restrictions must be adhered to. For details of the Copyright Agency licence contact:

Copyright Agency.
website: www.copyright.com.au
Email: info@copyright.com.au
Toll-free phone number (landlines only): 1800 066 844
Phone: +612 9394 7600
Fax: +612 9394 7601
Address: Level 11, 66 Goulburn Street, Sydney, 2000

This exciting new book *Primary Physics, the Principles behind Roman Machines* takes the reader on a short journey into the period in history when the Roman Empire was expanding, inventing, and inspiring changes which had never before been seen on such a scale. The Romans were interested in doing things bigger, better, and making things more aesthetically beautiful than ever before. Their legacy still stands many thousands of years later, in the form of the Colosseum, the aqueducts, and the many arches and monuments they left behind as testament to their achievements.

In this book, students and teachers will have an opportunity to experience the excitement and sense of accomplishment of taking on tasks that will challenge them and stretch what they think they can do. Building a working model of a battering ram or a rolling tower out of popsicle sticks, and rubber bands will stimulate the imagination and the joy of human simplicity. It's rewarding and it's fun. When I came up with the idea of making the blocks (that would become a Roman arch) out of plaster in ice cube trays, it made me laugh. When I actually created them, I was amazed by how beautiful and simple they were. And when it actually *worked* and the arch stood up by itself, I was overcome with joy and a sense of great satisfaction.

All the models are based on the working models in the Roman Machines Exhibition by Niccolai Teknoart SNC (Firenze). We have been working together with Luigi Rizzo, and his team to bring about what we think is a very excellent resource for learning. This book brings together the fundamental lessons of physics from our basic series *Primary Physics*, and links them together with models of the Roman machines. The lessons have been specifically adapted to sequentially build up an understanding of how physics principles apply to the operation of the Roman machines.

Acknowledgments

We would like to acknowledge the help we have received which has culminated in this book: Jan Dyers, herself a physicist, who had the concept that physics could and should be taught to young children; Ken Dyers, whose communication training program meant that we had the skills to be able to communicate simply about basic concepts. Ken's scientific approach to the subtle energies of communication expanded our reality on the application of modern physics to the human condition. His wisdom and the application of these subtle energies created a truly humanitarian science of communication. Also to the Kenja Social, Cultural and Sporting Association and the adults and children who participated in and supported the physics classes since they began there in 1985. We would like to thank Prof. Richard Collins for reviewing the book, his excellent suggestions, and his continuing support for the spirit of what we are doing.

M.E., A.D.

In memory of our friend Ken Dyers 1922–2007

DISCLAIMER

The publishers and authors have made every reasonable effort to ensure that the experiments and activities in this book are safe when conducted as instructed but assume no responsibility for any damage caused or sustained while performing the experiments of activities in this book. Parents, guardians and/or teachers should supervise young readers who undertake the experiments and activities in this book.

The content of this book is an integral part of the educational program of the Roman Machines Exhibition. Reproductions of work in this book are from the exhibition Roman Machines organised by the Artisans of Florence – Niccolai Teknoart SNC (Firenze). Used with permission.

a brief history of Rome

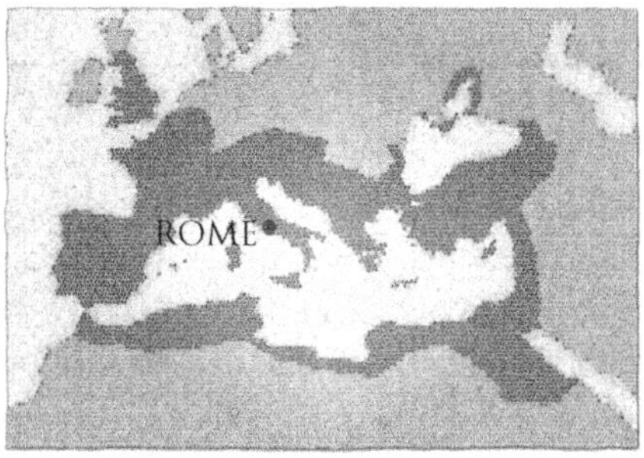

During the period from about 753 BCE to 476 CE, the Romans built a republic and an empire that survived for many centuries and brought many inspiring changes to civilization.

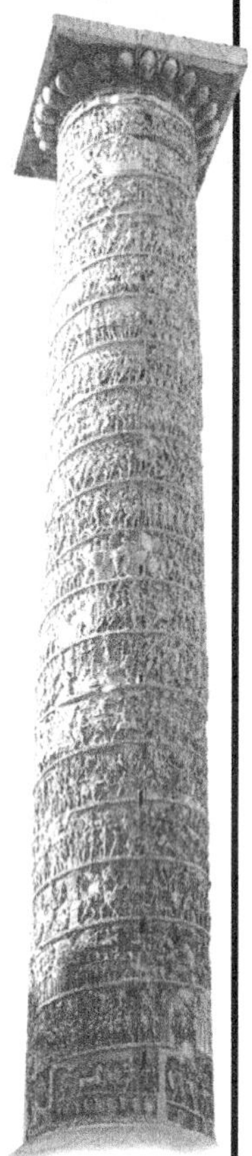

Much of what the Romans developed came from the people and the cultures that they conquered, particularly the Greeks. They picked out ideas that they wanted to use in order to build and run their empire.

The Romans respected and learned from scholars and incorporated many cultural ideas. Roman science was simple, practical, aesthetic, and always useful.

The Romans continually developed their cities. Many of their ideas still benefit us today, such as highways, public toilets, arenas, windows, concrete, calendars, books, calculators, notebooks, pocket timers, fast food and taxes!

The Roman leader, Julius Caesar (100 – 44 BCE) was a great military general. He modernized the calendar (the month of July is named after him) and under him great improvements were made to machines for building structures such as roads, arches, bridges and weapons of war.

JULIUS CAESAR

Julius Caesar also developed the idea of the book by dividing up scrolls onto separate pages that could be numbered. It was a military invention, brought about because of the need to refer to information quickly and easily in battle. Scrolls were cumbersome but a book allowed one to turn to a specific page and easily locate information.

The Romans used slaves, animals and gravity power to carry out heavy work, and to build the famous structures such as arenas, aqueducts, and monuments. Many of these still stand today.

Later they also used wind and water technology for producing goods, especially as the slave populations decreased. Places in Northern Europe and England where this occured became major centers of the Industrial Revolution in the 19th Century.

section one

writing and calculating instruments

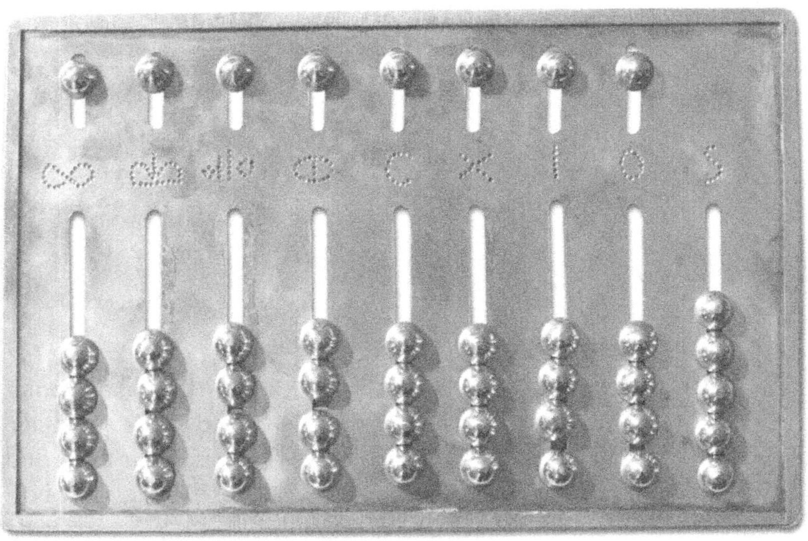

abacus

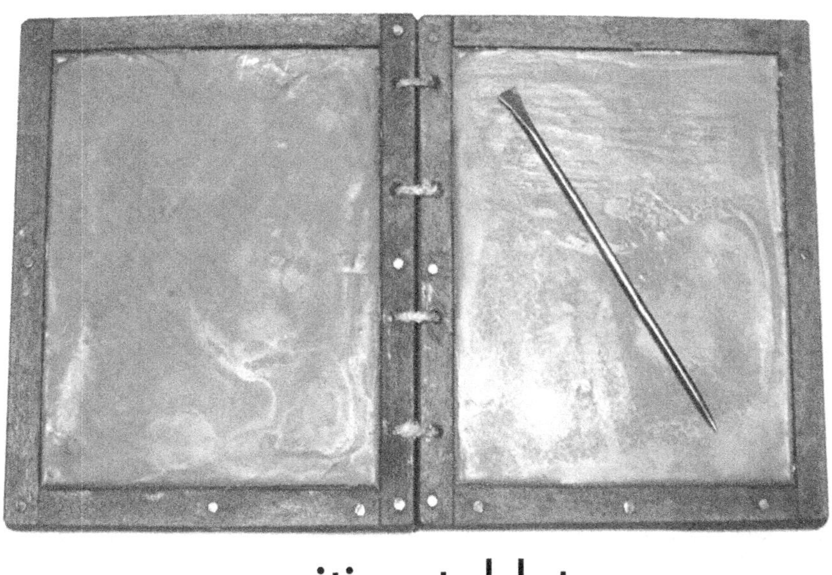

writing tablet

project 1 abacus

Materials

- 80 popsicle sticks (at least). Don't use any bent ones.
- 45 marbles, diameter 15 mm (5/8")
- craft glue
- shears or scissors (**important**: for adult use only)
- wooden board, at least 25 cm x 25 cm; surface must be very flat

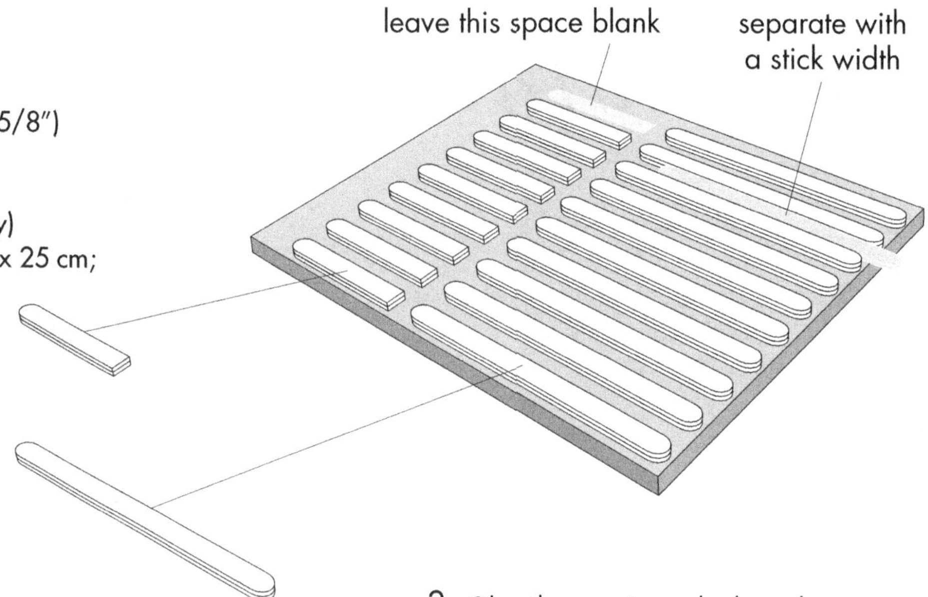

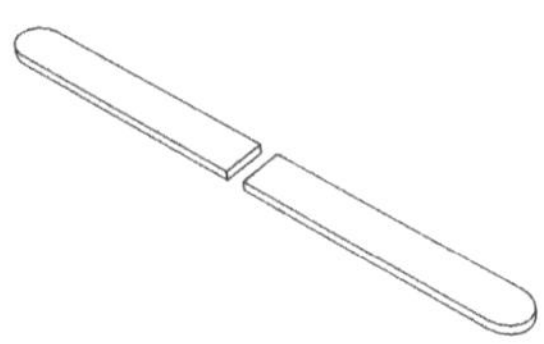

1 Cut 14 popsicle sticks in half.

2 Join 9 half sticks to 9 half sticks and 10 full sticks to 10 full sticks.

3 Glue these pairs to the board as shown. Use another stick to make sure the spacing is even and parallel. Note there is one less pair of half sticks than full sticks.

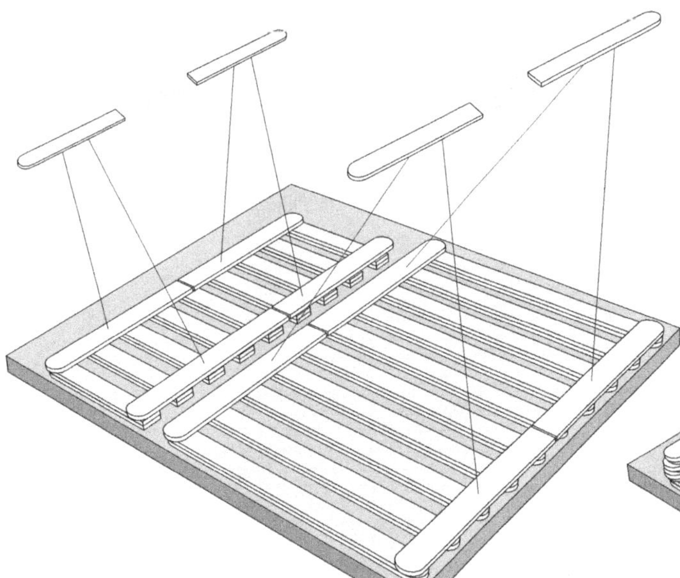

4 Ask an adult to cut 4 sticks with the shears or scissors so they are the size to fit neatly with two across the top and two across the bottom ends of the row of sticks as shown.

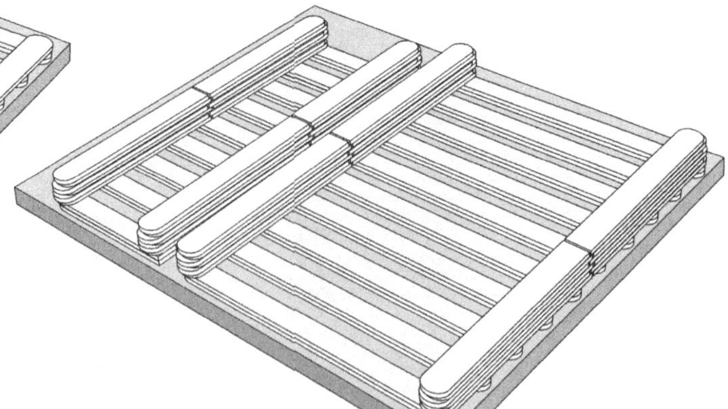

5 Add 5 more layers of sticks of the same lengths as the previous step.

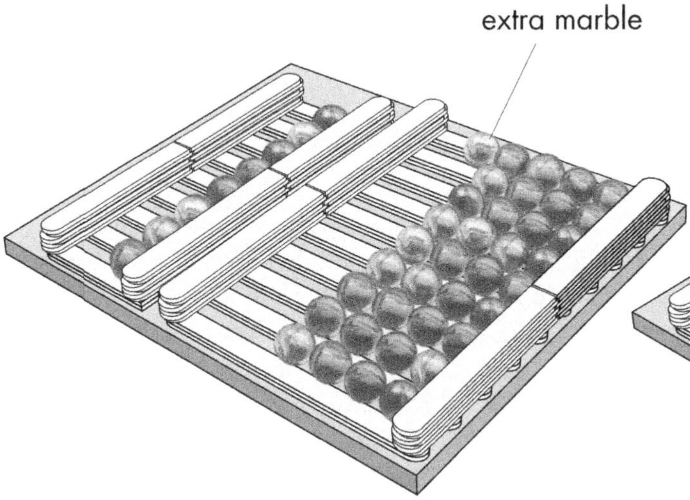

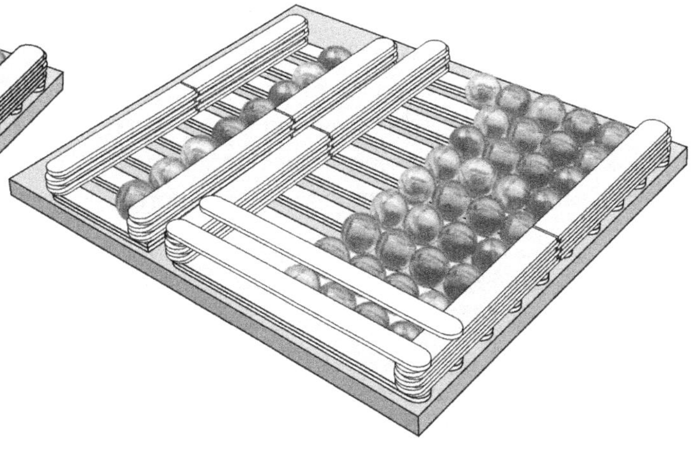

6 Place one marble in each half-stick slot and 4 marbles in each full stick slot. Place an extra marble in the last slot.

7 Glue 2 full sticks above the first slot. Check whether the marbles fit snuggly, i.e. they can move easily, but not roll around. If it is too tight you may have to add more layers of sticks, or even glue a layer of cardboard to achieve the right height.

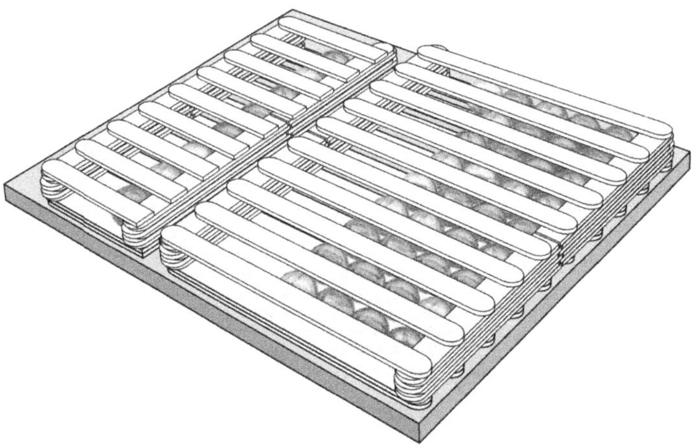

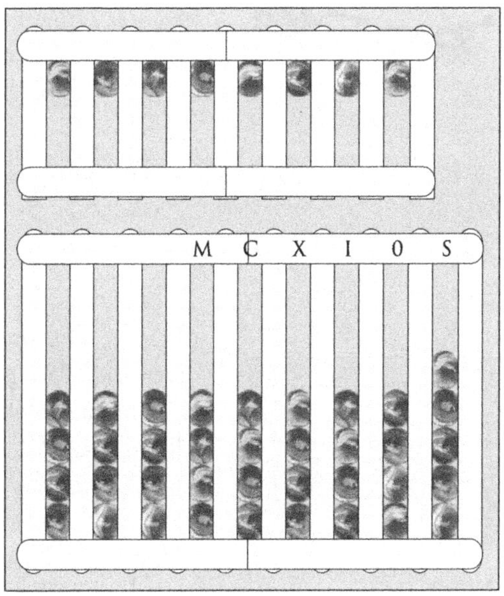

8 Add the remaining rows at the same height.

9 Inscribe numbers on each slot to indicate their value.

abacus

The ancient Romans used an abacus to help them with their counting and calculating of numbers. The Roman abacus was made up of columns of beads or pebbles (in this model we use marbles).

There are two rows of columns with a bar in between them. Each column signifies a different decimal value. The columns on top are short and contain one marble each signifying "five". The columns below are longer and contain four marbles each signifying "one". The one furthest to the right has five.

The two columns on the right are for calculating fractions. The last row has five beads so it can calculate twelfths (rather than tenths). The columns to the left are "I" which is the "ones" column; "X" the "tens" column; then the "hundreds" titled "C"; and then a series with symbols that signify the "thousands", "ten thousands", "hundred thousands", and "millions" columns.

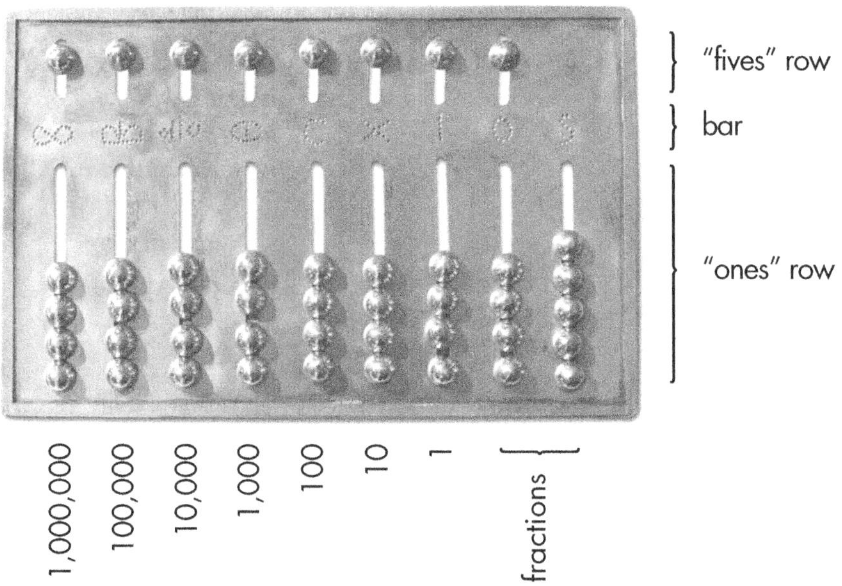

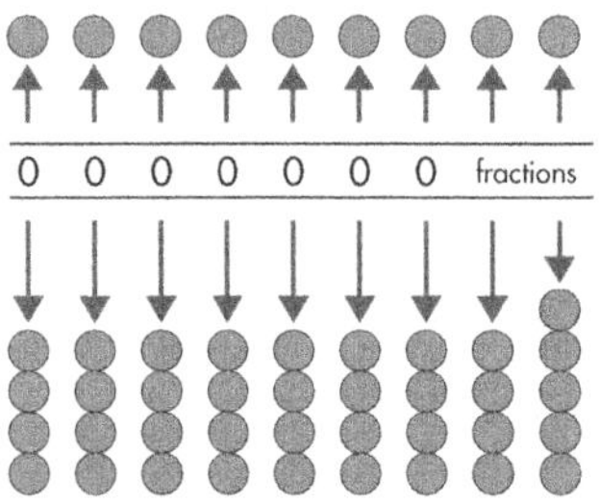

To begin using the abacus, you must first clear the abacus. This is done by pushing all the marbles in the direction away from the bar between the columns, i.e, push the marbles on top up, and the marbles in the lower columns down.

To show a number, you push the correct number of marbles towards the bar. You must always move from left to right.

For example, to show 637, move a "five" and a "one" in the 100s column; 3 "ones" in the 10s column; and a "five" and 2 "ones" in the 1s column.

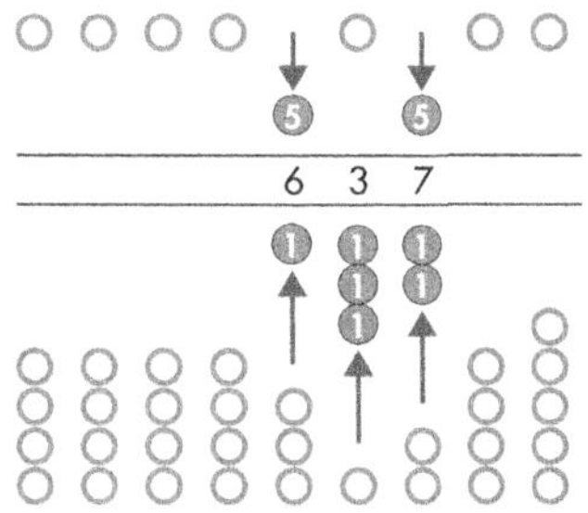

This will be shown simply as:

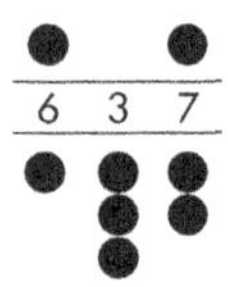

abacus addition

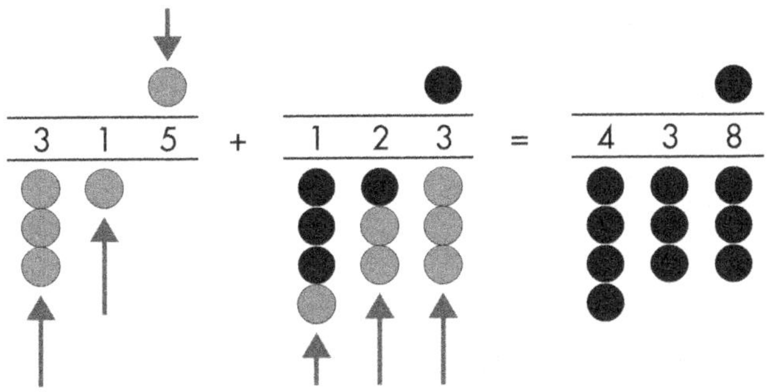

315 + 123

Set up 315: move 3 "ones" in the 100s column; a "one" in the 10s column; and a "five" in the 1s column.

To add 123: move a "one" in the 100s; 2 "ones" in the 10s; and 3 "ones" in the 1s column.

Thus 315 + 123 = 438

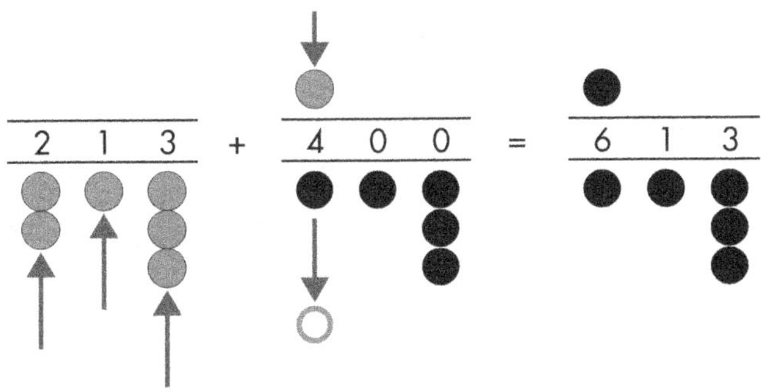

213 + 400

Set up 213: move 2 "ones" in the 100s; a "one" in the 10s; and 3 "ones" in the 1s column.

To add 400: there aren't 4 "ones" left to move so we move a "five" in the 100s, then take away a "one" in the 100s column. There are no changes needed to 10s and 1s columns.

Thus 213 + 400 = 613

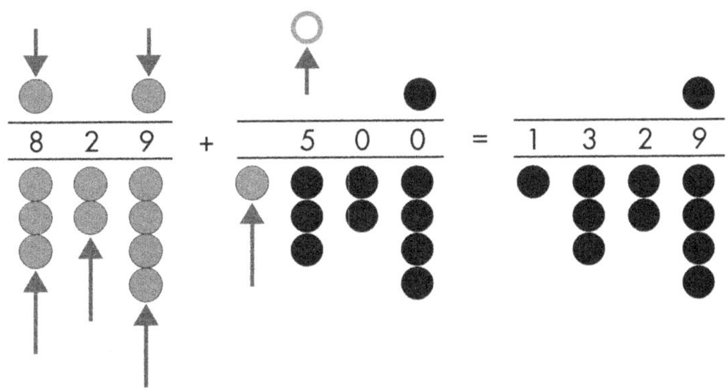

829 + 500

Set up 829: move 3 "ones" and a "five" in the 100s column; 2 "ones" in the 10s; and a "five" and 4 "ones" in the 1s column.

To add 500: there is already a "five" in the 100s column, so we move it back then add a "one" in the 1,000s column.

Thus 829 + 500 = 1,329

Try these sums on your abacus:

231 + 340 =

782 + 131 =

1,317 + 422 =

22,456 + 34,403 =

abacus subtraction

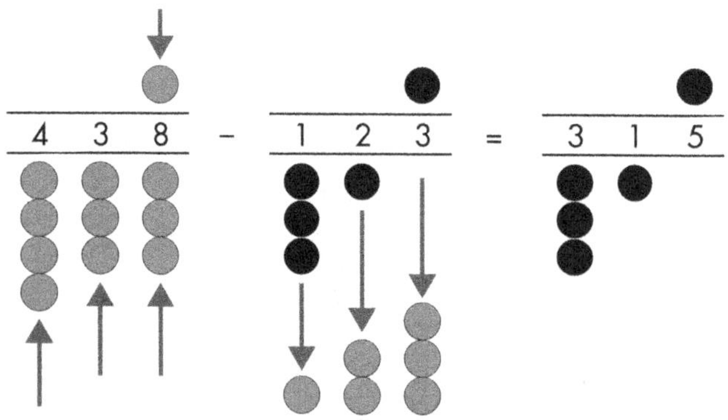

438 – 123

Set up 438: move 4 "ones" in the 100s column; 3 "ones" in the 10s column; and a "five" and 3 "ones" in the 1s column.

To subtract 123: move *away* a "one" in the 100s; 2 "ones" in the 10s; and 3 "ones" in the 1s column.

Thus 438 – 123 = 315

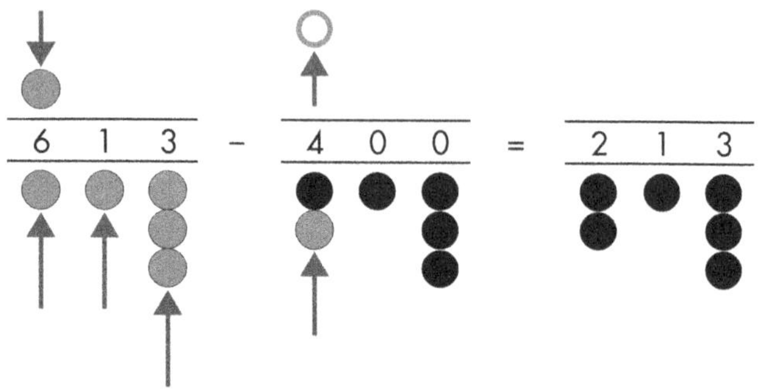

613 – 400

Set up 613: move a "five" and a "one" in the 100s; a "one" in the 10s; and 3 "ones" in the 1s column.

To subtract 400: we move away a "five" in the 100s, then *add* a "one" in the 100s column (so we use 400 = 500 – 100).

Thus 613 – 400 = 213

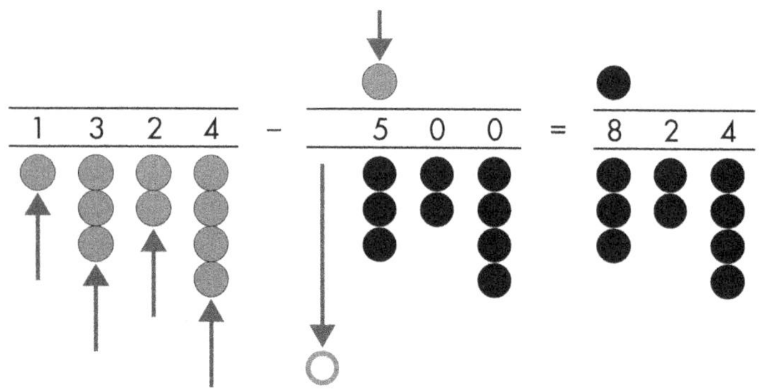

1,324 – 500

Set up 1,324: move a "one" in the 1000s; 3 "ones" in the 100s column; 2 "ones" in the 10s; and 4 "ones" in the 1s column.

To subtract 500: we take way the "one" in the 1,000s column and add a "five" in the 100s column (using 500 = 1000 – 500).

Thus 1,324 – 500 = 824

Try these subtractions on your abacus

641 – 320 =

782 – 131 =

1,617 – 422 =

28,456 – 34,403 =

wax tablet

Wooden tablet with wax (tabulae ceratae)

From the earliest schools in Rome, young students acquired their first notions of writing by practicing on wooden tablets topped with a layer of wax.

Writing on sheets of papyrus or parchment was very expensive so the wax tablets were common not only in schools, but were used extensively in everyday situations, as a sort of notebook.

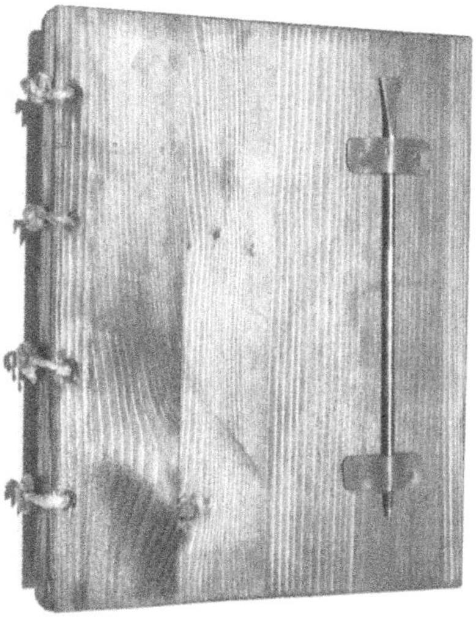

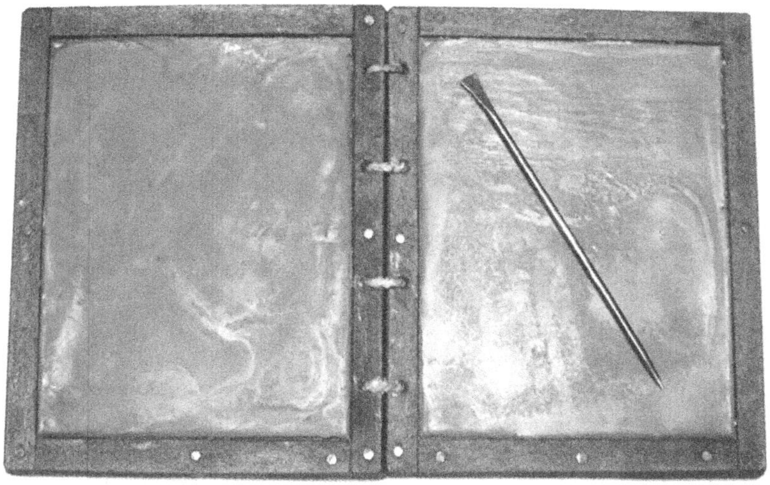

The writing was done by means of a metal strip with a tip on one side (stylus or graphicum), which was used to etch the wax leaving marks of writing.

The other end of the strip was slightly flat and served, after warming it on a candle, to clear writing by accurately smoothing the tablet.

Woman using her stylus and book.
From a Pompeii fresco

project 2 wax tablet

Materials

- whiteboard (or blackboard) with wooden frame, approx. 20 x 30 cm (8″ x 12″)
- Plasticine or beeswax, approx. 500 g (1 lb). Note: beeswax is more expensive though more authentic.
- rolling pin or glass jar
- waxed/baking paper
- BBQ skewer
- pencil

1 Cut a piece of baking paper about 60 cm (2′) long and place it on a flat table.

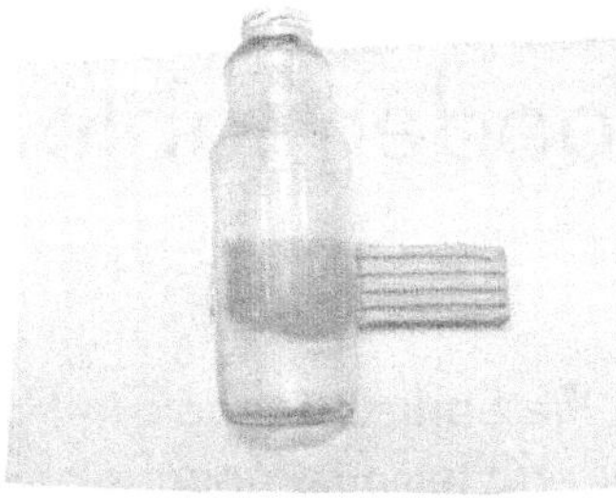

2 Take a slab of plasticine and roll it out flat on the paper with the bottle or rolling pin.

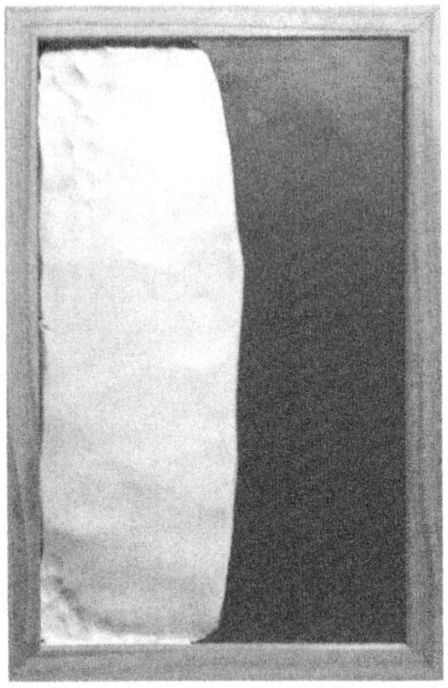

3 When it is smooth and approx. $1/2$ cm ($3/8$″) thick, peel it off the paper and place it within the frame of your board.

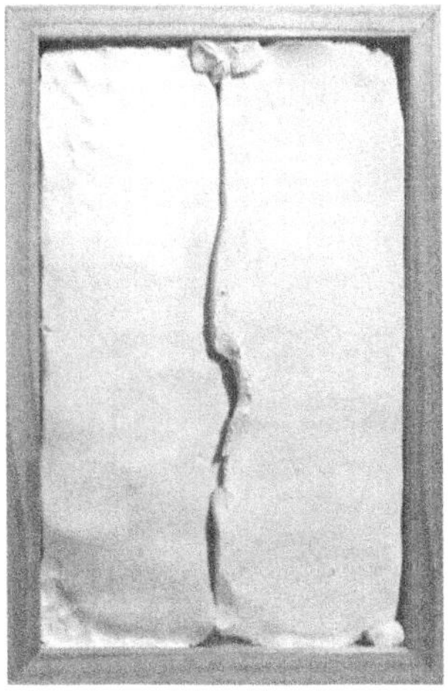

4 Continue to roll pieces and place them in the frame until you have covered the inside of the frame with a thin layer of plasticine.

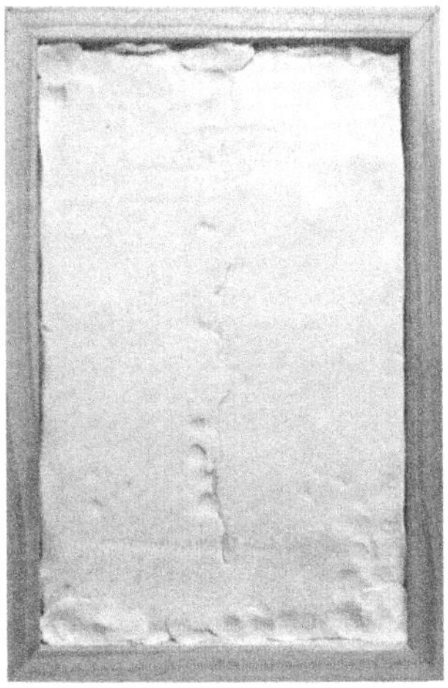

5 Press to seal any seams and mold it into the corners.

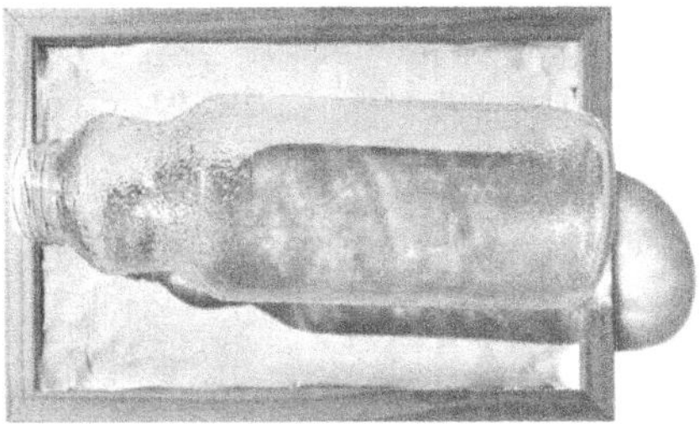

6 Roll again with the bottle to make it even.

7 Use the skewer to write.

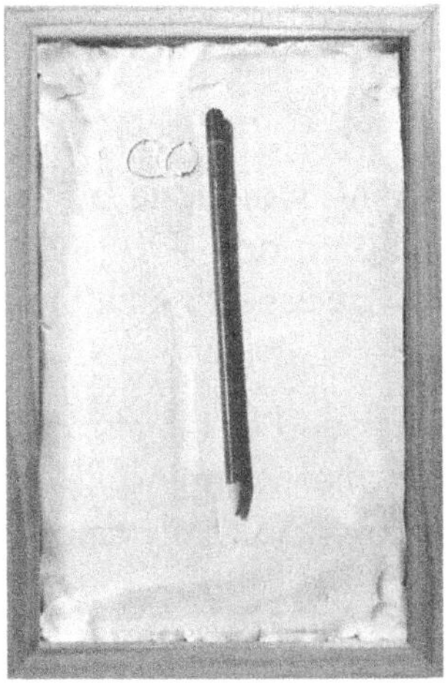

8 To erase, use a round pen or pencil like you would use the rolling pin.

section two

Roman roads

project 3 Roman roads

Materials

- 1 large rectangular take-away container, approx. 10 cm x 20 cm (4″ x 8″)
- 1–1.5 cups large pebbles (approx 1–2 cm)
- 1–1.5 cups coarse gravel OR marble chips, approx. 0.5 cm ($^{1}/_{4}$″)
- 1–1.5 cups dry sand
- 1–1.5 cups smooth stones (approx. 3–5 cm (1″–2″)
- plaster (OR glue made from flour and water)
- water (to mix plaster as per directions on packet)
- plastic container to mix the plaster or glue
- popsicle stick
- rubber mallet

1 Fill the bottom of the container with a layer of large pebbles approx. 2–3 cm (1″) high. Flatten them out as much as possible by gently shaking the container horizontally and gently tapping with your hand.

2 Put a layer about 1.5 to 2 cm ($^{1}/_{2}$″) of coarse gravel on top. Gently shake it to flatten the layer allowing pieces to fall into the spaces between the pebbles.

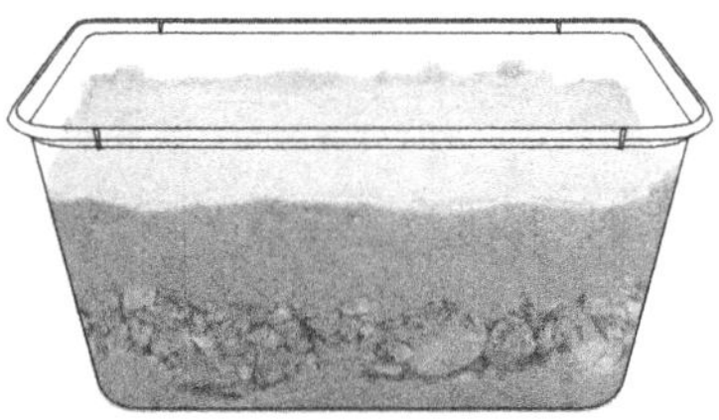

3 Pour a layer of sand about 1.5 to 2 cm ($^{1}/_{2}$″) on top of the gravel. Shake gently, allowing sand to fill the spaces.

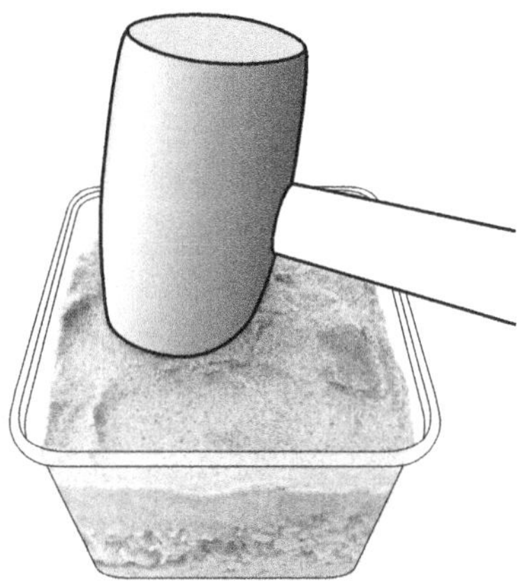

4 Gently pound the sand with a rubber mallet until it is flat and compressed.

5 Place a layer of smooth stones on top. Fit them together as best you can to cover the sand. Gently press them into the sand so they stay in place.

6 Press down so that the stone surface is slightly higher in the middle, sloping downward towards the long edges of the container.

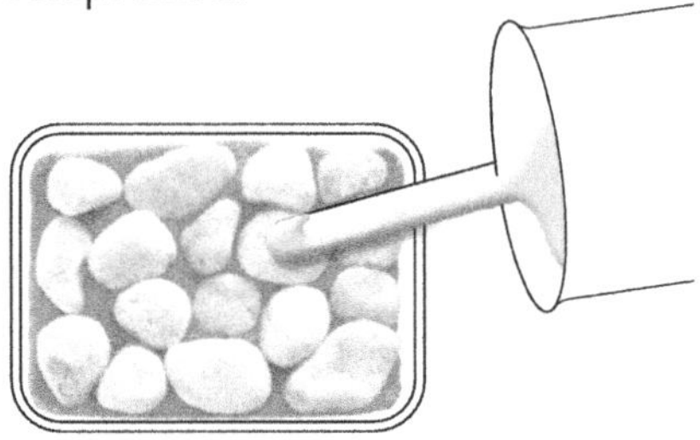

7 Mix the plaster and pour carefully into the spaces between the stones, filling all the spaces where you see sand, including all around the edges. Holding a popsicle stick to the side of the plaster's container can help direct the liquid plaster where you want it to go.

8 Let it set. You now have a model of the way the Romans constructed their roads, some of which have lasted thousands of years. Now we are going to investigate *why* this is such an effective design.

mass

Physical things

Place all of the objects on a table.
Examine each (e.g. lift it, squeeze it, scratch it with your nails.)

Which materials do you think would be best for building a road? Why?

____

____

Which would not be suitable for building a road? Why?

____

____

Things you need

- rocks of various composition and size, e.g. quartz, pumice, etc
- solid objects of different composition, e.g. wooden blocks, plastic balls, rubber balls

Words to use

shape size
heavy light
texture mass

Extension

- Make a chart of the different types of mass you examined. List their properties.

mass	1.	2.	3.

e.g.
1. squeezable,
2. can scratch with fingernail,
3. cracks when dropped

space

Things you need

- lots of empty drink cans, cartons, milk cartons, empty toilet rolls OR blocks of *same* size
- poster boards in shapes

Words to use

shape	pack
stack	fill
circle	triangle
square	rectangle
hexagon	octagon

Extension

- Ask students to fold up newspaper "logs" using sticky tape. Let them build large structures like forts and towers, big enough for them to fit inside.

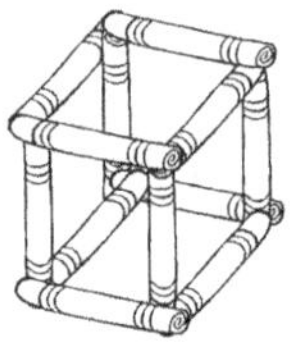

- Make patterns using piles of coins.

Mass takes up space

Fill up the poster board shape with your containers, blocks or cans.

What happens to the flat shape when you fill it in with cans?

____

Do the cans fill the whole area or are there some spaces between the cans?

____

Do the rocks and pebbles in your model road fill the whole space?

____

What did you fill the spaces with? __________________

What did the Romans use to fill up the spaces between the rocks in their roads? ______________

____

volume

Things you need

- small container of known volume (e.g. 250 ml / half pint container of juice)
- larger container which is 2, 3 or 4 times the volume of the small containers (e.g. 500 ml / pint, 1 litre / quart) container of juice)
- lots of marbles, small blocks, plastic cubes
- measuring cup

Words to use

fill	full
empty	pour
spill	hold
more	less

Extension

- Fill a small container with marbles. Count how many it holds. Then fill a larger container with marbles. Count them. Compare the two results.

  What is the relationship between them?

Filling space

Volume is a measure of the space occupied by any material, or the size of a container that can hold that material.

Fill a small container with mass, e.g. marbles or small blocks.

Pour them into a larger container.

Fill the smaller container again and repeat until the large container is full.

How many times do you fill the small container in order to fill the large container?

What is the total volume of rocks, sand, gravel, etc that you used to make your model section of the road?

Calculate the volume as follows:

Measure the height of rocks, sand and gravel.

Mark a line on an empty container.

Fill with water to line.

Tip contents into measuring cup.

Write the volume above.

The volume of water is equal to the volume of your model Roman road.

gravity and mass 1

Things you need
- wooden board
- flat head nail
- very flexible rubber band OR elastic string
- 3 equal rubber bands
- paper clip
- plastic container OR plastic cup (punch 3 holes with a hole punch in advance if possible)
- any objects of known weight, e.g. 50 g, 100 g, 200 g (2 oz, 4 oz, 8 oz)
- scissors OR hole puncher to **carefully** make holes in the cup (Note: Adult supervision is important)
- hammer

Words to use

mass	gravity
pull	exactly
more	less
between	close to

Extension

Use different combinations of masses to further calibrate the gravity machine.
(See next page.)

Gravity causes the force of attraction between two masses, e.g. a ball and the earth.

Give the cup a pull so that the rubber bands are taut. Then let go before marking the 'zero' line.

Make something to measure the pull of gravity.

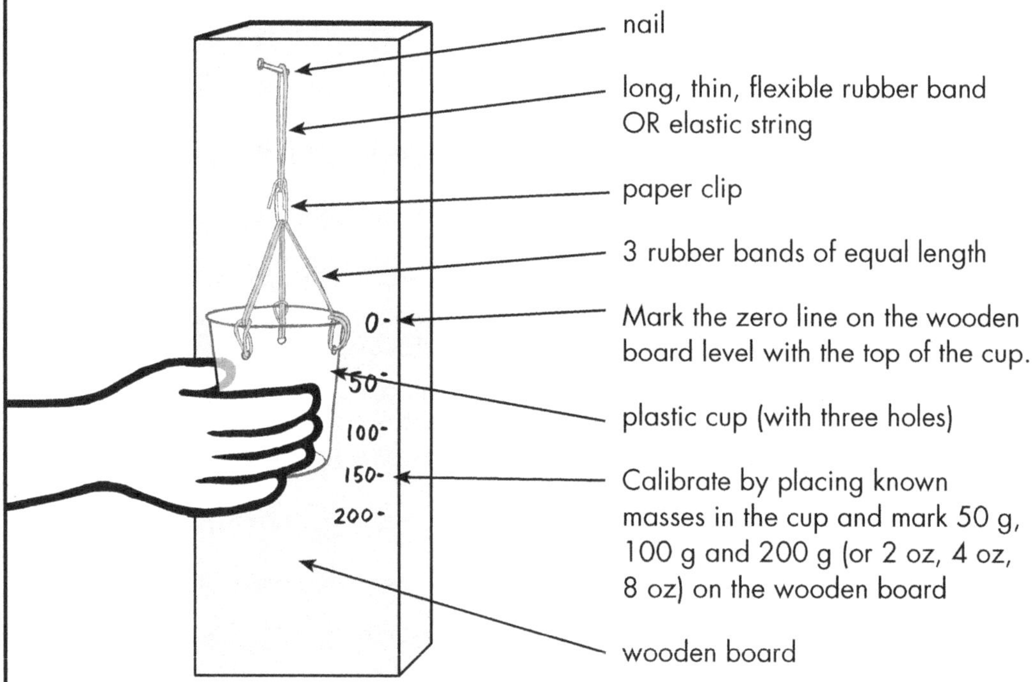

What force makes your machine work? ________________

gravity and **mass** 2

Things you need

- gravity machine from previous lesson
- golf ball
- marbles
- potato
- hard-boiled egg
- onion
- apple
- orange

Words to use

mass	bigger
smaller	scale

Try your machine with different amounts of mass.

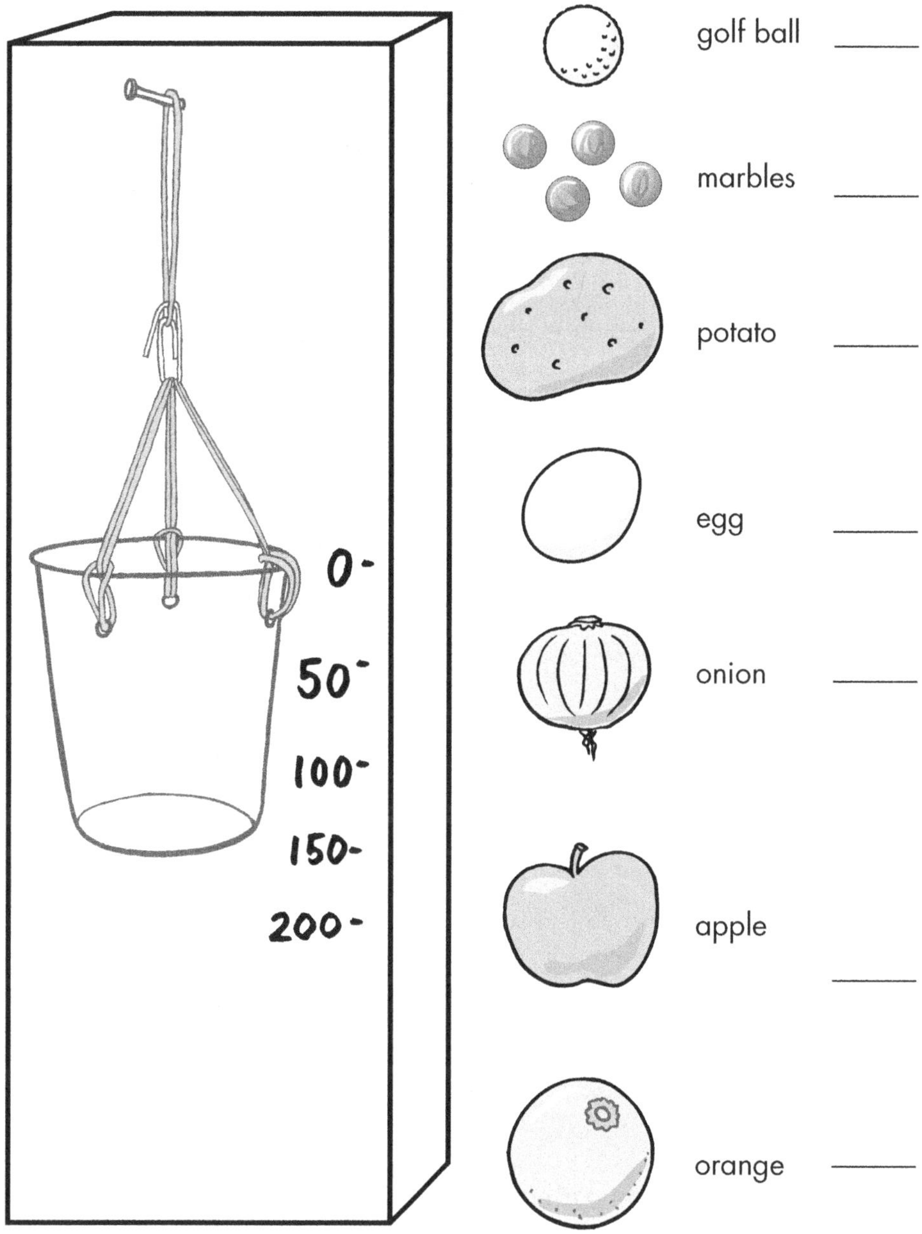

Draw a line from each object to the point it measures on the scale. Write the approximate mass on the line next to the picture.

Extension

How many marbles does it take to make 100 g (4 oz)?

gravity and **mass** 3

Things you need

- 2 identical small containers e.g. empty 250 ml (8 oz) fruit cups
- sugar
- sand
- a ruler and a pencil to make a scale

Words to use

more mass
less mass
level balance
equal same
weight density

Extension

Which way must you move the pencil so that the two masses balance?

At what number on the ruler does it balance?

Try measuring both masses on your gravity machine from the last experiment.

Same size, different mass

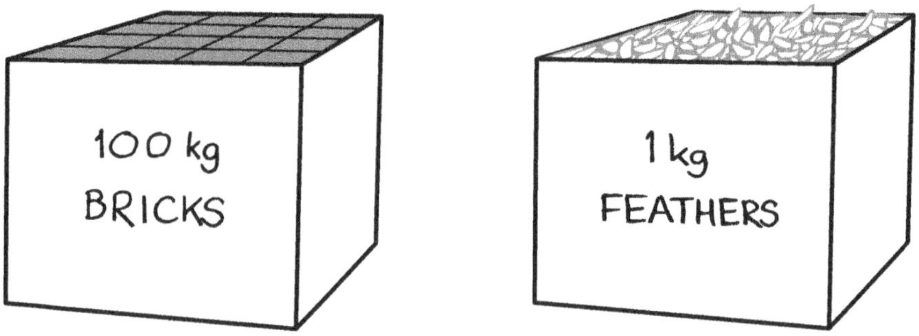

Fill the two containers. Level the top with a ruler.
Put them on the scale to see which has more mass.

Which one has the most mass? ______________________

Can two things have the same size but a different mass? ______________________

gravity and **mass** 4

Things you need

- kitchen scale
- large plastic container
- rice
- water
- marbles
- coins
- other uniform objects

Words to use

pull	mass
gravity	force
strong	soft
how much	weight
heavy	light

Extension

Locate a scale at home or in a shop. Measure the weight of things at home or in the shop.

Scales measure weight

A scale measures the amount of gravity pulling on the mass. This measurement is the object's **weight**.

Fill containers to the same volume as your Roman road sample. Find their weight using a kitchen scale.

Fill in the chart:

object	weight
My sample of a Roman road	
Same volume of rice	
Same volume of water	

Which has the greatest mass? ____________________

density

Things you need

- paper
- aluminium foil

both cut to same size

Words to use

more	less
packed	space
size	large

Extension

- Try wetting the paper to see if it can be packed more tightly.
- Compare the number of used paper plates or paper cups that can fit in a garbage bag:

  1) if they are just thrown away individually;

  2) if they are stacked before throwing away.

  Which is more dense?

  If you were paying the garbage collectors per bag, which method would you use?

Packing in

How small can you make a piece of paper?

How small can you make the same sized piece of aluminium foil?

Which one can be packed the most tightly or **densely**? ____________________

Describe what is happening to the paper and aluminium foil.

____

When you filled up the spaces in your sample Roman road, how did that affect its density? ____________

____

Here is the formula for density: **density** = **weight** ÷ **volume**

weight from page 21 volume from page 17

What is the density of your sample Roman road? ____________

Now we understand why the Romans' road design works so well. The different sized stones and sand form into a high density material that doesn't break up easily. The strength of modern day concrete is based on exactly the same principle.

section three
Roman arch

Roman arch

I'm sitting on top of one of the greatest inventions in history – THE ARCH.

My fellow countrymen used arches in the design of our buildings and developed magnificent structures based on the humble arch.

When made of lightweight material, like our newly invented concrete, multi-storey buildings became possible to build.

Our engineers were inspired and designed monuments, bridges, aqueducts and many other things using arches – practical things which will stand the test of time. The importance of the arch is that it increases space and volume, while reducing mass and weight.

The parts of an arch are:

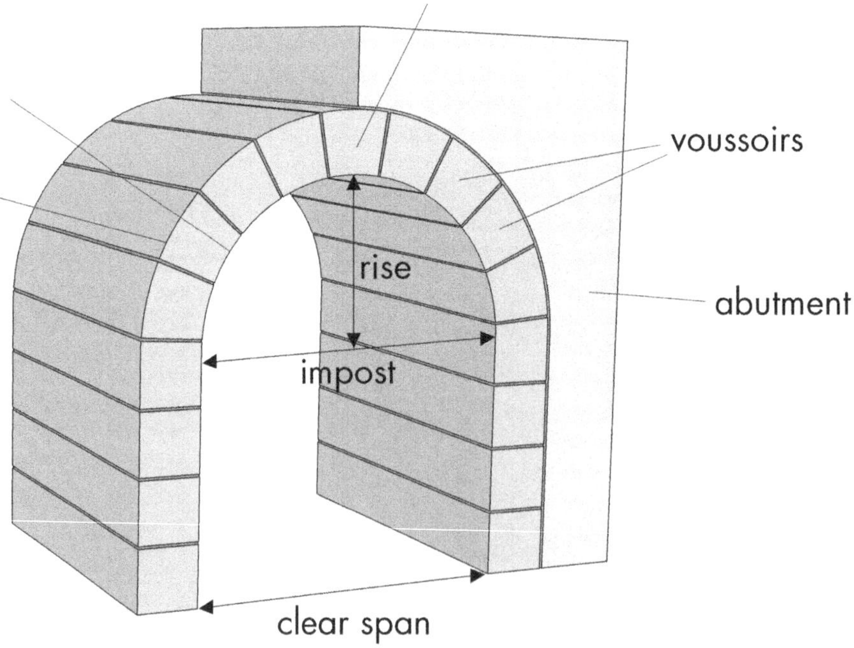

project 4 Roman arch

Materials

- ice cube tray
- plaster
- water
- plastic container to mix
- popsicle stick (to stir, and scrape off plaster powder level)
- sand paper (rough and smoother)
- masonite placemat
- piece of cardboard
- pen
- 15 cm (6") ruler

1 Mix plaster according to instructions. Measure carefully so the plaster sets properly. Scrape the powder flat on the spoon to get accurate quantities.

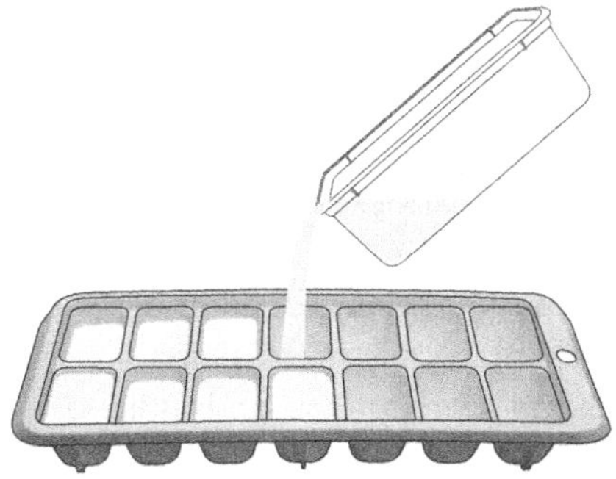

2 Quickly but carefully pour the liquid plaster into an ice cube tray. Fill the cubes to the top, and all the same size. Make about 12 blocks.

3 Allow to set. Tap the tray on a table a few times to get any bubbles to rise to the surface.

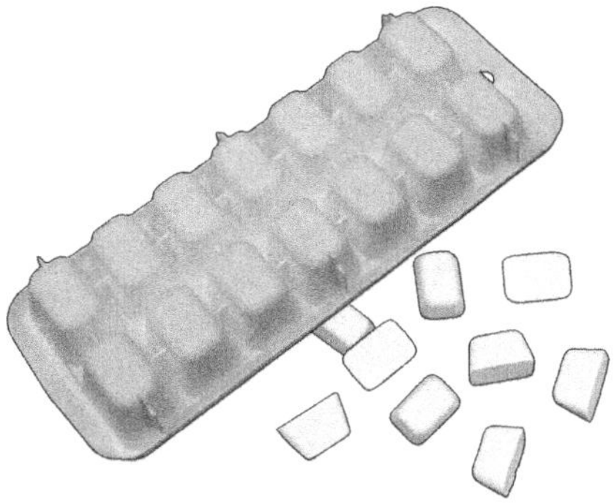

4 When hard, remove from tray.

5 Allow to dry overnight, or longer, depending on weather conditions.

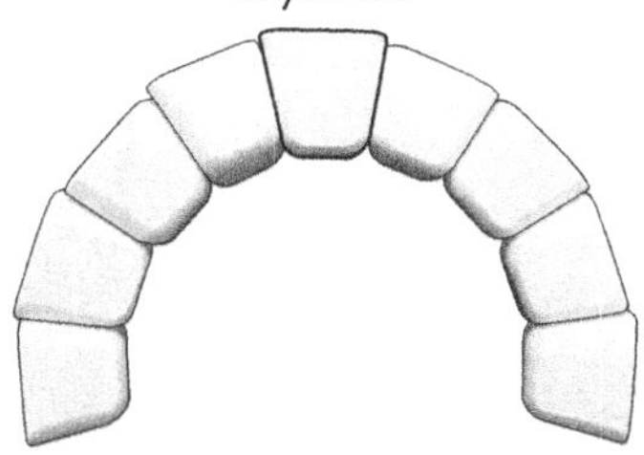

6 Working on a flat surface, lay the blocks down with the largest block you have as the keystone, (top center). Line up several blocks, side by side to make an arch shape.

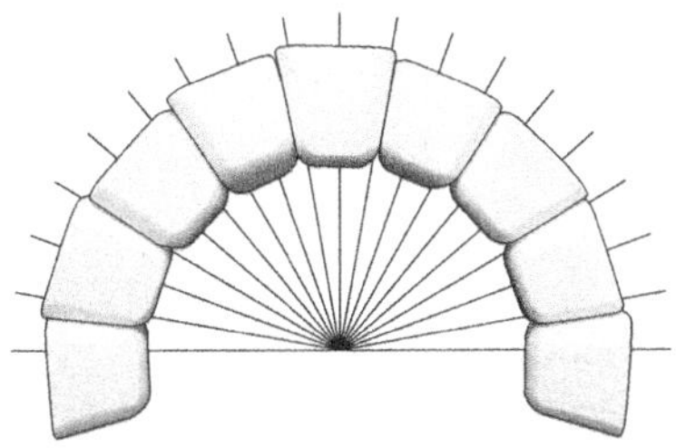

7 Depending on the shape of the blocks, you may need 7, 9 or 11 blocks to make an arch of at least 180°.

8 Rough up the 4 sides of each block with the smooth sandpaper. This will help them hold onto each other. Be sure to only sand in one direction, and don't change the shape of the blocks.

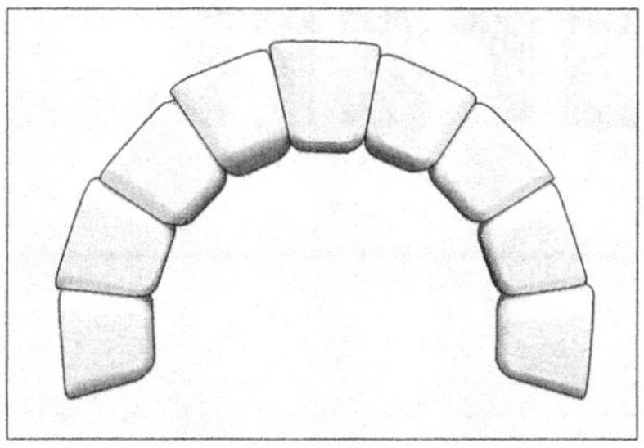

9 On the placemat, assemble the blocks back in the arch shape. Make sure they are pushed as tightly together as possible, while maintaining shape, and that the arch is lined up straight on the mat.

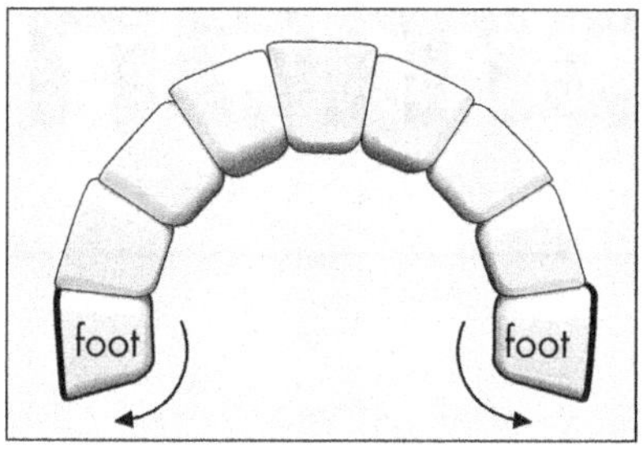

10 Rotate each foot a quarter turn.

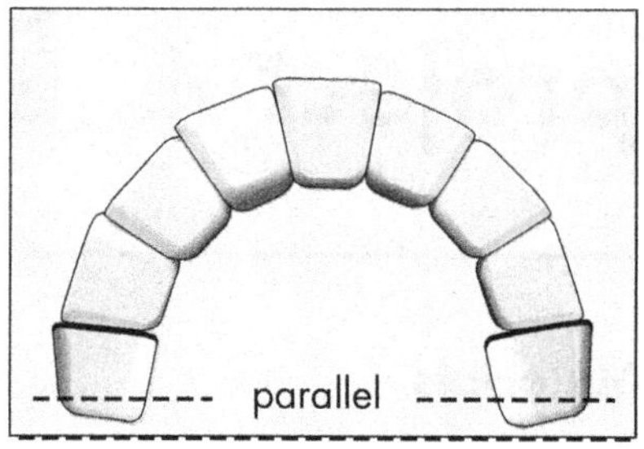

11 Draw a line with a pen across the bottom of the foot, at an angle parallel to the bottom edge of the placemat. Use a small ruler.

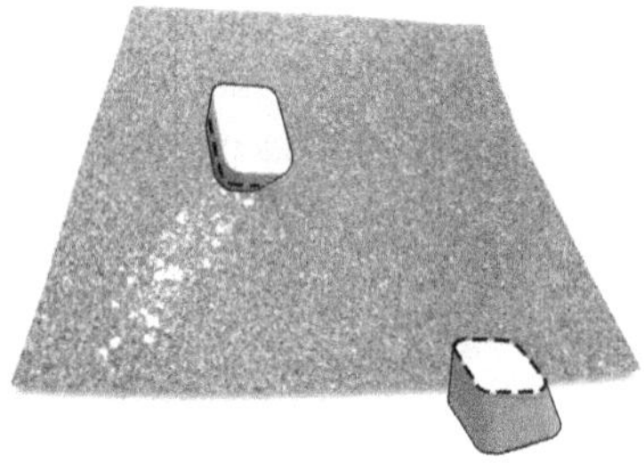

12 Use rough sandpaper to sand off the plaster up to each line. You will now have two blocks that fit into the angle of the arch, yet are parallel to the ground. (Mark them 'left' and 'right' as they might differ slightly.)

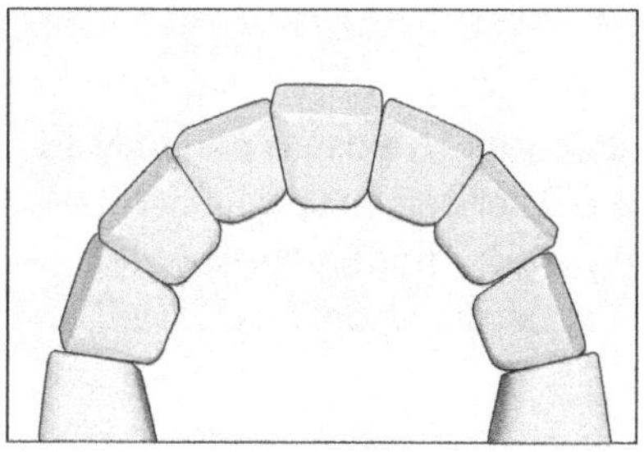

13 Carefully slide all the blocks down the placemat so the two feet are even with the bottom edge. Make sure all the blocks are sitting snuggly next to each other.

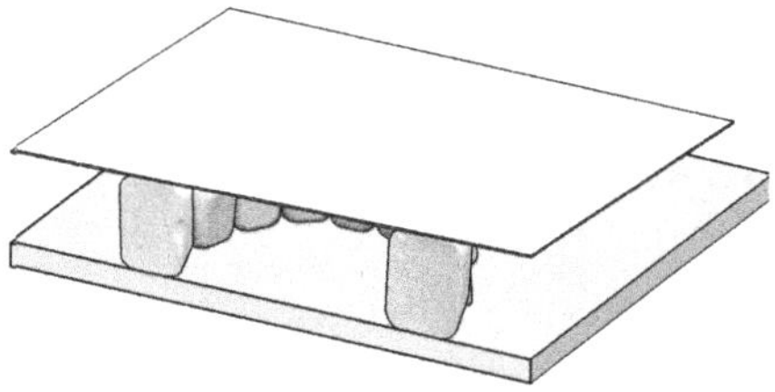

14 Place the piece of cardboard over the arch. Hold one hand on top.

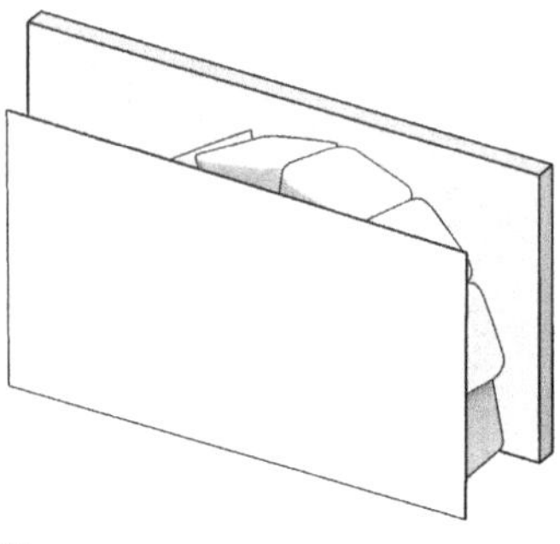

15 Carefully with the other hand, tilt the placemat so it is perpendicular to the table and the arch is sandwiched between the placemat and cardboard.

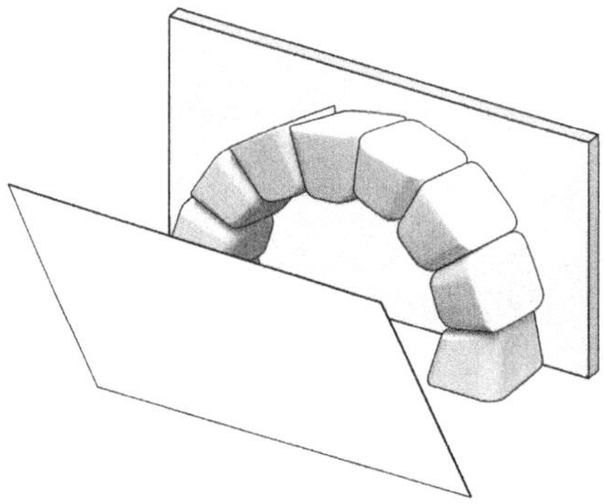

16 Remove the cardboard slowly, and carefully adjust any blocks which have moved out of alignment.

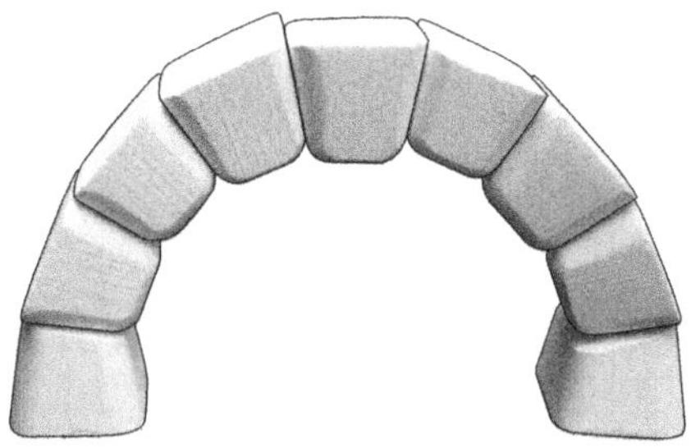

17 Remove placemat from behind. Your arch is standing!

Extension: Try building another arch out of blocks where you have added tiny styrofoam balls (like from a bean bag), to the ice cube tray before adding the plaster. Use these lightweight blocks at the top of the arch. How does it compare to the first model? The Romans built many structures with lighter materials on top, and heavy materials on the bottom for stability. Some arches and some of the famous domes were constructed in this way.

forces

Things you need

- ball of soft dough
- table

Words to use

friction	rubbing
gravity	falling
mass	force
move	still
energy	cause
push	pull

Extension

- Try moving other objects along the table to feel how much friction they make. Watch gravity at work as they fall off the table.

Mass

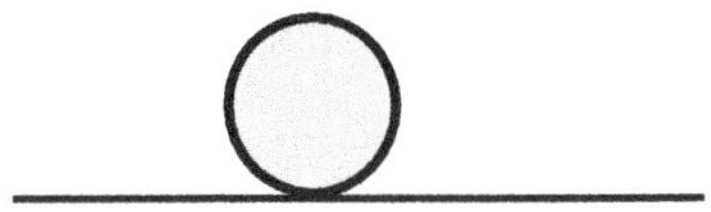

Force used on mass

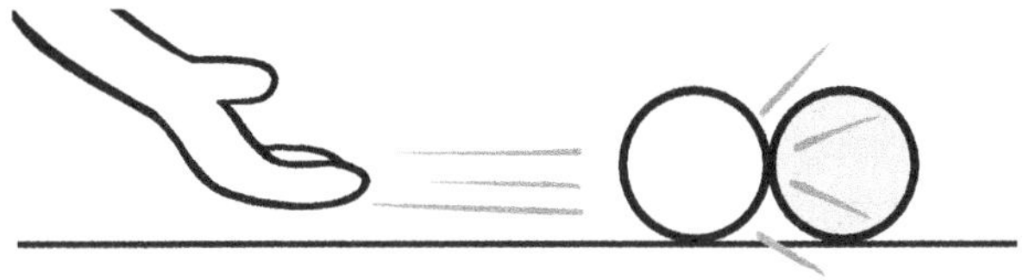

1. Place a ball of dough on a table.
Quietly watch it for a while. Don't move. Does the dough move?

What does it do? ____________________

2. Push it with your finger. It moves! (Important: don't let it roll; slide it.)

What force can you feel? ____________

3. Slide it so it drops off the edge.

What force can you now see acting? ___

We call the rubbing force **friction**.

We call the falling force **gravity**.

In your Roman arch, how is friction helping to hold it up?

____

How is gravity helping to hold it up? ________________________

____

friction

Things you need

Different surfaces:

- paper
- sandpaper, different grades 6 cm x 8 cm (3″ x 5″) each
- waxed paper
- recycled paper
- cardboard
- chalk
- crayon
- soft pencils

Words to use

rough smooth
rub friction
heat small
disappears

Extension

See how long it takes to wear away these:

1 cm (1/2″) of chalk on sandpaper;

1 cm (1/2″) of crayon on sandpaper.

Note: Use a longer piece so you can hold it safely. Measure 1 cm from the end and mark that spot.

Rubbing

Draw on different types of surfaces

When you rub on sandpaper, what happens to the crayon, chalk or pencil? ______________________

Is the sandpaper still as rough after you have used it? ________

What happened? ______________________

Friction is the rubbing force. The rougher the surface, the **more** friction there is.

How did you use friction to form your block? ____________

In your Roman arch, identify all the places where there is friction. ______________________

How is friction helpful? ______________________

How is it harmful? ______________________

What changes could you make in your Roman arch to take advantage of friction? __________

____

The Romans used arches such as this in many situations, including bridges, windows and aqueducts. The same method is still used today.

section four
pile driver

project 5 pile driver

Materials

- 22 popsicle sticks
- string, approx. 40 cm (16″)
- coins, around 12 mm (1/2″) diameter
- masking tape, 25 mm (1″) wide
- plastic drinking straw (cut in half)
- glue
- bolt
- 2 washers
- 2 nuts
- sewing bobbin
- shoebox

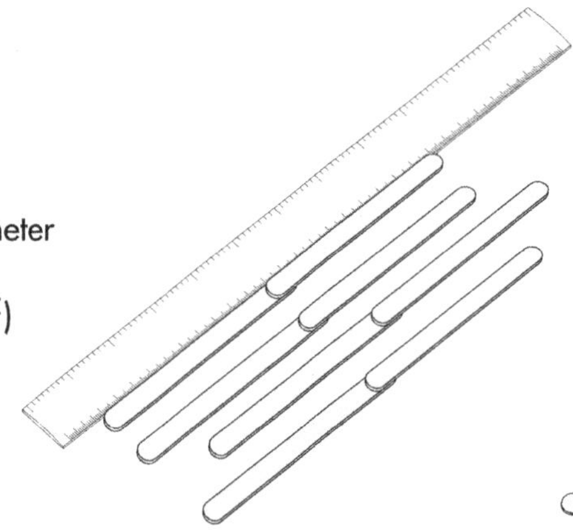

1 Glue two sticks together in a straight line. Push them up against a ruler to make sure they are straight. Repeat this so you have four double length pieces.

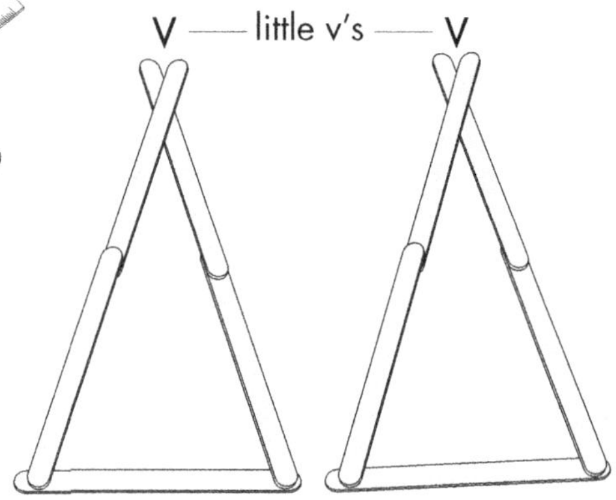

2 Glue the sticks together as shown, making two triangles with a "little v" on top. Be sure that the ends of the long sticks do not poke out past the base stick.

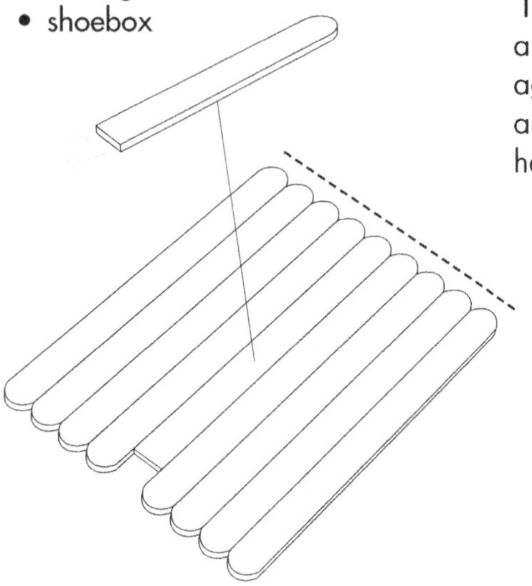

3 Cut the end off one popsicle stick. Glue 4 sticks together to both sides of this stick, keeping them as even as possible along the base.

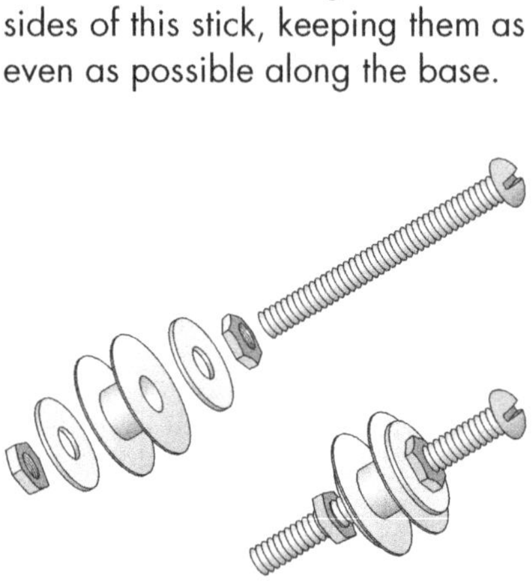

4 Put the washers, bobbin and nuts on the bolt as shown.

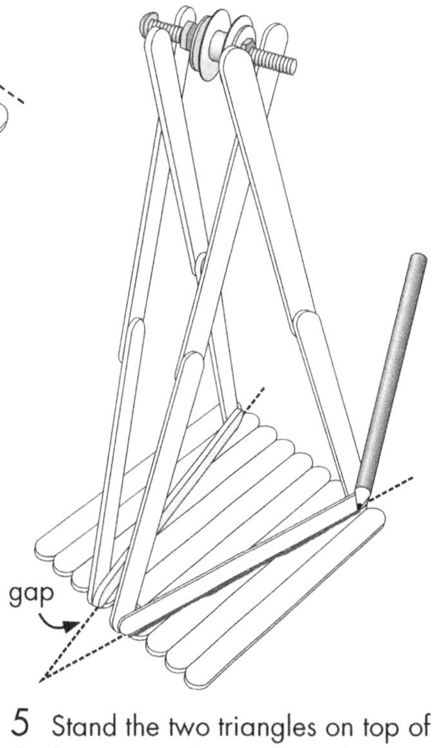

5 Stand the two triangles on top of the base on a diagonal, as shown. Be sure they are fixed in position outside of the gap, i.e. they do not overlap it. Set the bolt in the two little v's on top. You may want to draw lines with a pencil on the base to show where the triangles will be glued.

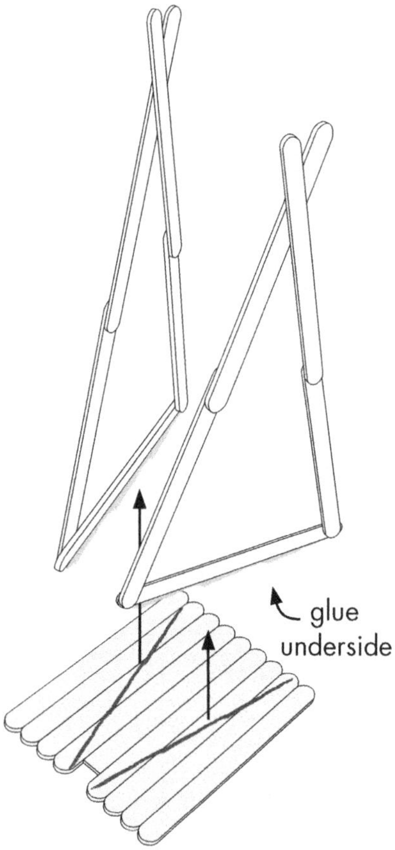

6 Lift up your pieces, and put glue along the narrow edges of the base popsicle sticks.

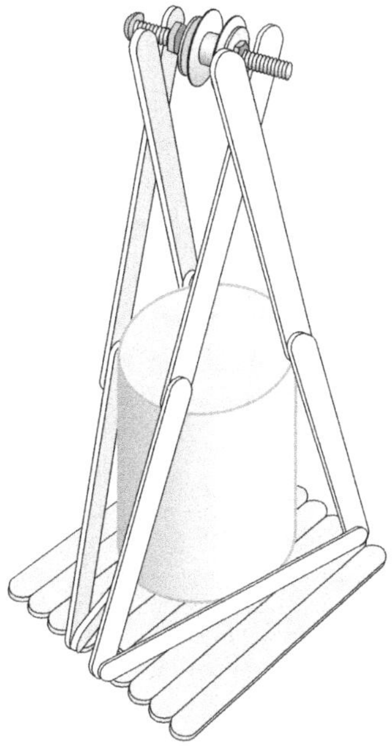

7 Stick them down on the lines you've just drawn. Put the bolt back in the two little v's to hold the triangles in the correct position. Place some small rather heavy objects (like a small tub of fruit) up against the triangles to hold them in place while the glue dries.

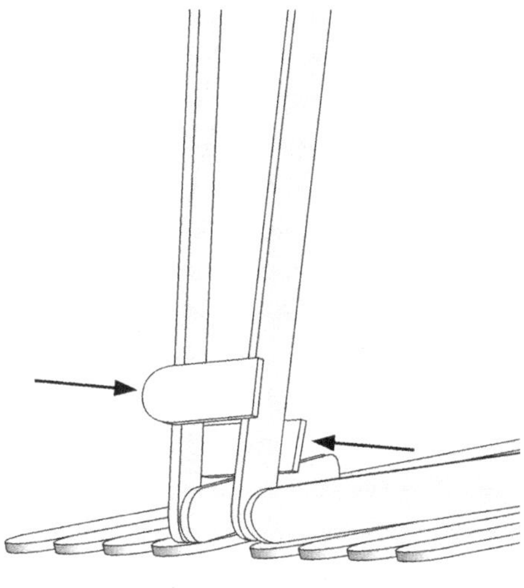

8 Glue two quarter popsicle sticks on the uprights at the front as shown.

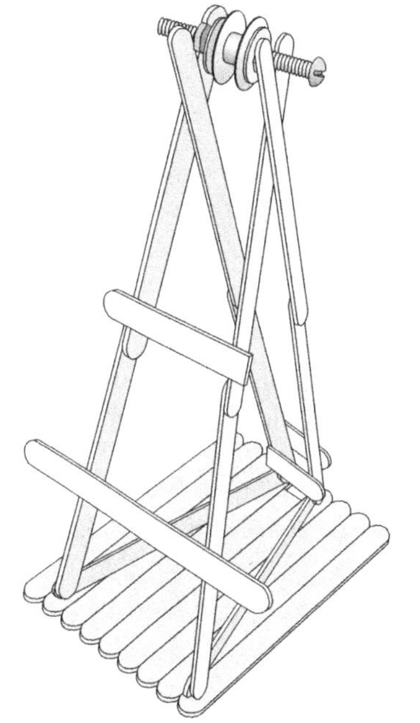

9 Glue a popsicle stick $1/3$ of the way up from the base, along the open side of the "big V". Glue another $1/2$ popsicle stick $1/3$ of the way down from the top of the open side of the "big V".

10 Make a pile of 6 to 8 coins.

11 Wrap them up in masking tape. This is your weight.

12 Tie one end of the string around the weight. Wrap it up securely and tie a knot.

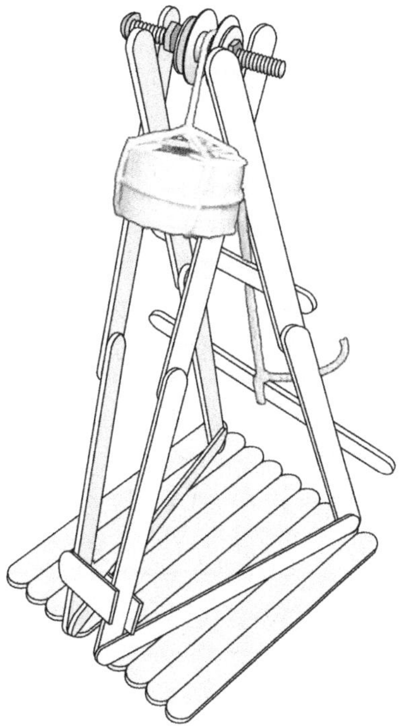

13 Drape the string over the bobbin, pull the weight up to the top, and secure the free end of the string by tying it onto the lower cross piece.

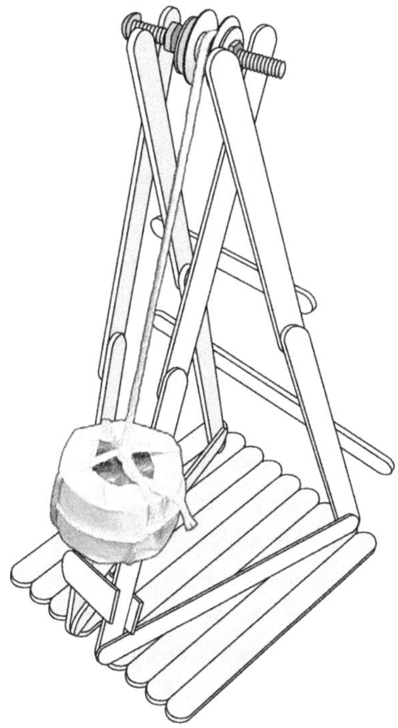

14 Check that the weight drops freely when you release the string. (It is this falling heavy mass which drives the pole into the ground.)

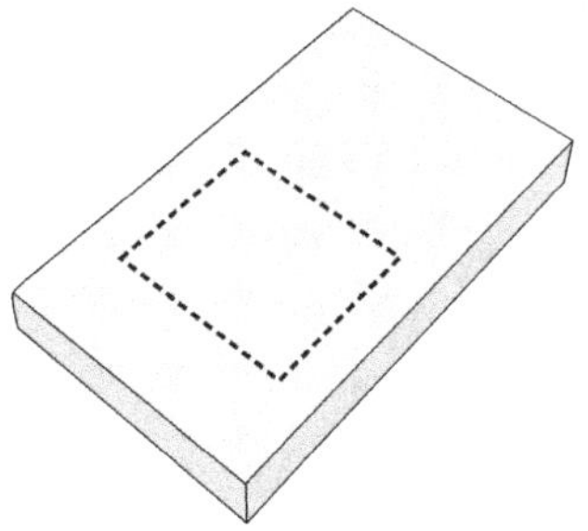

15 Cut a hole in a shoebox lid as shown with the dotted line.

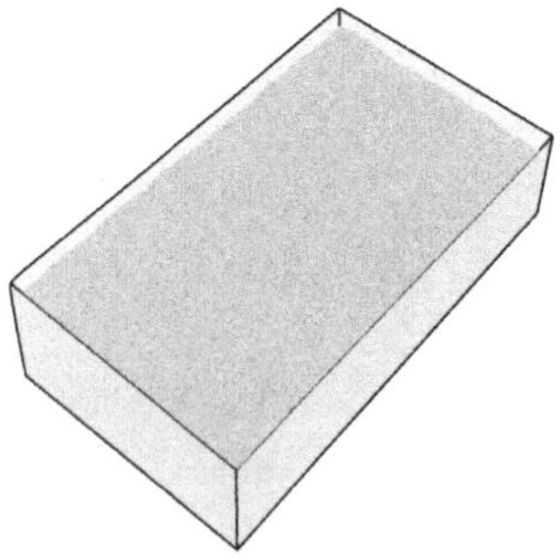

16 Fill the shoebox with sand.

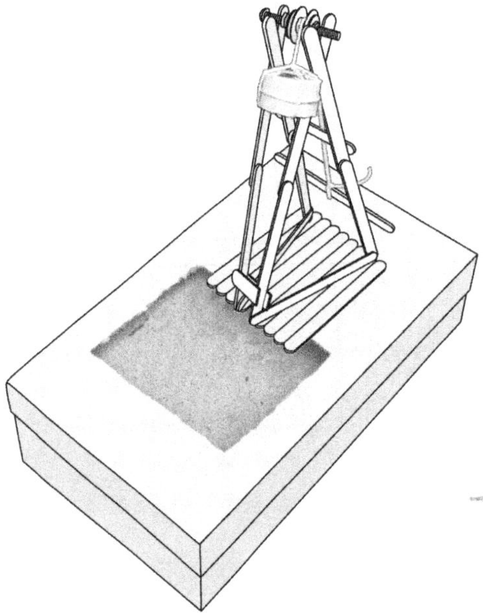

17 Put the lid on the shoebox then position the pile driver on top.

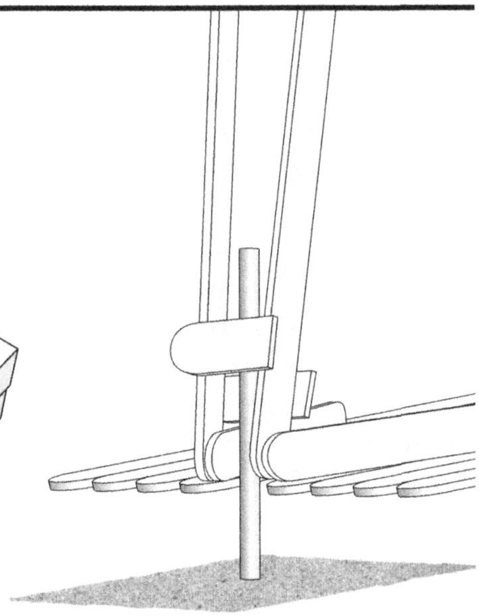

18 Position the straw so it is vertical inside the channel above the gap in the platform. Anchor one end in the sand.

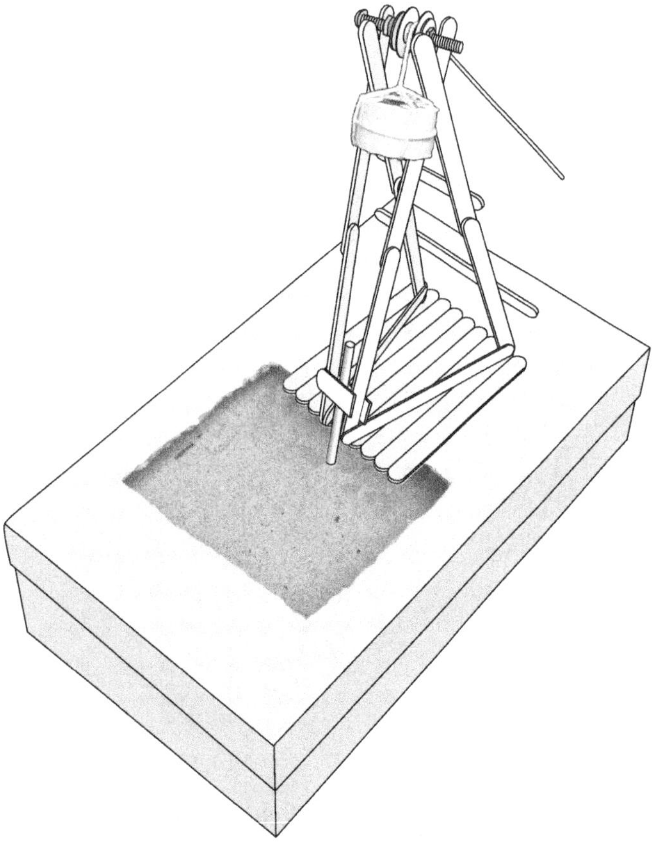

19 Untie the string. Make sure the weight is positioned properly at the top of the channel so it will slide down when you let go.

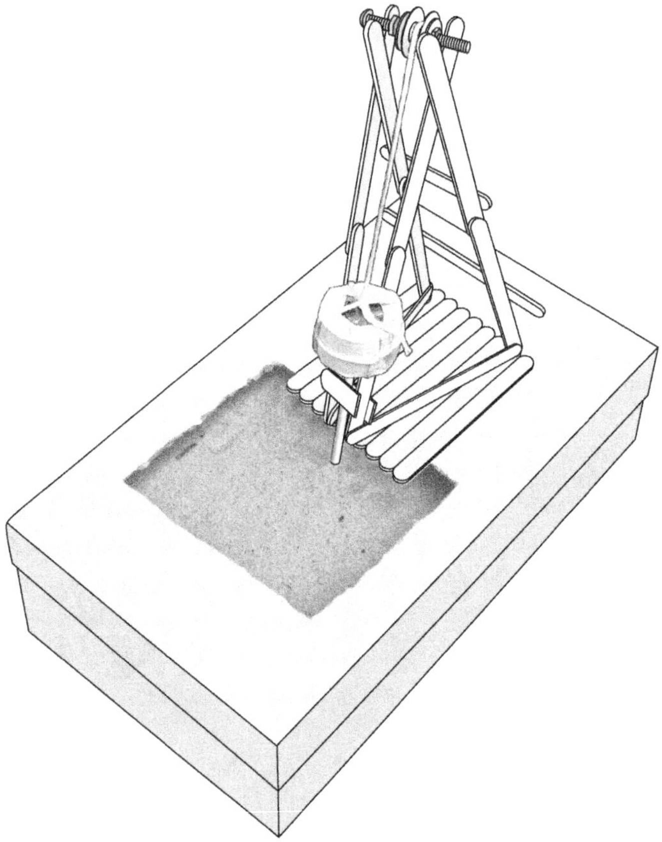

20 Let go. The mass will fall and pound the straw into the sand.

gravity 1

Things you need

- soft ball of dough
- smooth floor

Words to use

fall, drop, gravity, distance, compare, difference, force, energy, size

Extension

- Drop a hard-boiled egg (in its shell) from a height of about 1 cm (1/2 inch). Increase the height from which you drop it by 1 cm at a time. At what height does the shell crack?

Falling

1. Drop a piece of soft dough from a low height.

What happens to its shape?

____

2. Drop it from up high. **Be careful not to fall.**

What happens to its shape?

____

Why does this happen?

____

When you lift up the mass with the pulley of your pile driver, you are working against this 'falling force' or **gravity**.

When you lift the mass up and then drop the mass so it hits the top of the pole, what happens to the pole? Why?

____

____

pulleys 1

Things you need
- bucket
- pulley
- short rope
- long rope
- water (or weights)
- climbing bars (or monkey bars)

Words to use

raise	lower
force	friction
gravity	higher

Be careful to stand behind the line.

Extension
Have relay races lifting the bucket up to the top a certain number of times without spilling the water.

Excursion
Indoor rock climbing:

Measure how high you climb by measuring the length of rope pulled.

Note how many pulleys are used.

A **pulley** is another use for a wheel and axle.

Set up a pulley. Use a climbing bar, rope, pulley and bucket filled halfway with water.

Draw a safety line on the ground. Everyone has to stay behind the safety line until their turn.

Note: **Be careful! No one should be underneath the bucket.**

Try lifting the bucket up to the bar using just your arm.
Can you do it? _________

Now use the pulley to raise the bucket to the bar.

Can you do it? _________

Is it hard work? _________

Give everyone a chance to lift the bucket up as high as possible 5 times in a row using the pulley.

When you pull downwards on the pulley, which way does the bucket move? _____

Safety note: When lowering the bucket, *slowly* move your hands "hand over hand".

pulleys 2

Things you need

- 3 pulleys
- 2 ropes, 3 m (3 yards) each
- 2 buckets
- water (OR weights)
- short pieces of rope (to tie the pulleys to the railing and bucket)

Words to use

pull	longer
force	shorter
fixed	further
work	moveable
easier	harder

Extension

Where do you see pulleys used in your home?

What are they used for?

Draw a picture of your pulleys.

Try this

Tie a rope to one broom handle and wrap the rope around both several times. Hold on to the free end. Ask your friends to pull the sticks apart, while you pull on the rope. They won't be able to. You've made a very efficient pulley system.

Set up a single and a double pulley.

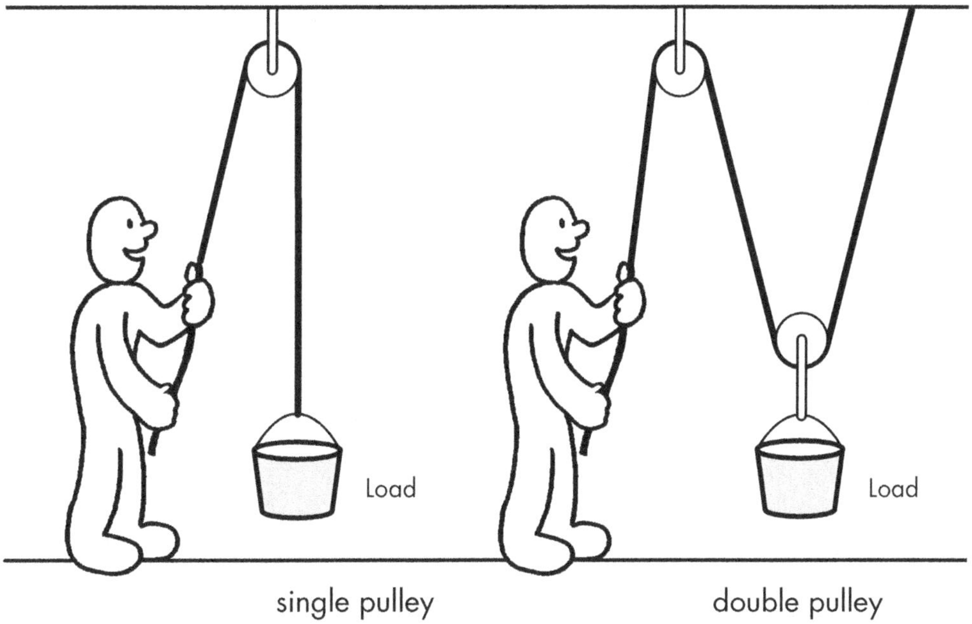

Put the same amount of water (or weight) into each bucket.

Have a turn lifting each bucket.

Label the fixed pulleys. Label the moveable pulleys.

Which bucket is easier to lift? ______________________

Do you think that the Romans would have used double pulleys in their pile drivers? Why? ___________

____

What are some advantages of using a double pulley? ______

____

What are some disadvantages of using a double pulley? ____

____

pulleys 3

Things you need

- single and double pulleys (from the previous lesson. Set up in advance if possible.)
- pencil (to copy down data)
- tape measure
- marking pen

Note

To measure the length the rope was pulled:

1. Pull the rope so that the bottom of the bucket is at the designated height. Make sure the rope is taut and vertical. Mark the rope clearly where it touches the floor.

Take care to stand aside

2. Lower the bucket to the ground and again mark the vertical rope where it touches the floor.

3. Measure the distance between the dots.

Measure how far you need to pull the rope to lift the bucket.

Single pulley

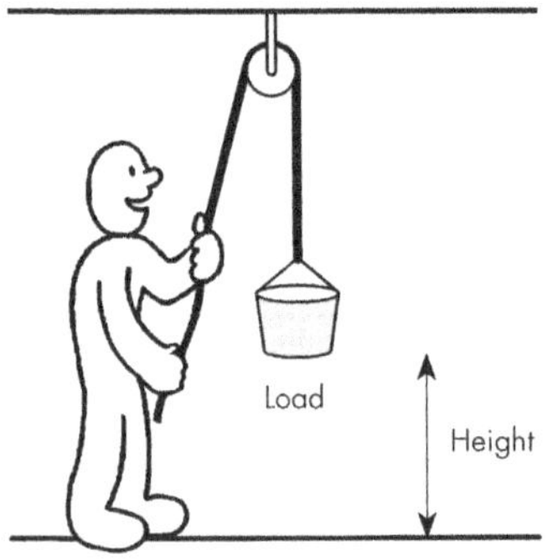

Height bucket pulled	Length rope pulled
30 cm (1′)	
60 cm (2′)	
90 cm (3′)	

Double pulley

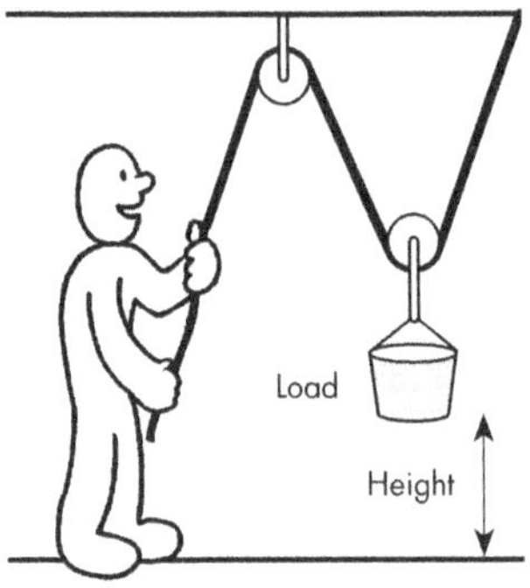

Height bucket pulled	Length rope pulled
30 cm (1′)	
60 cm (2′)	
90 cm (3′)	

What pattern did you see for the single pulley? ______________________

____

What pattern did you see for the double pulley? ______________________

____

What conclusion can you draw from these results? ______________________

____

If Vitruvius* was in battle and you were his assistant, and he needed to pull a large, heavy object up quickly, what kind of pulley would you recommend and why? (Note: Remember the Romans had slaves and animals to help them carry out heavy work.) ______________________

____

* For information on Vitruvius see back cover.

section five

battering ram

project 6 battering ram

Materials

- 29 popsicle sticks
- craft glue
- 1 dowel, approximately 2 cm (1″) diameter and 28 cm (12″) long
- plasticine or modeling clay (a lump approximately 2 cm (1″) diameter)
- 30 cm (12″) length of twine
- 8 rubber bands, size 14 (or similar, just not too thick)
- a toy animal head (a fridge magnet is good as it has a flat back)
- 10 empty toilet rolls

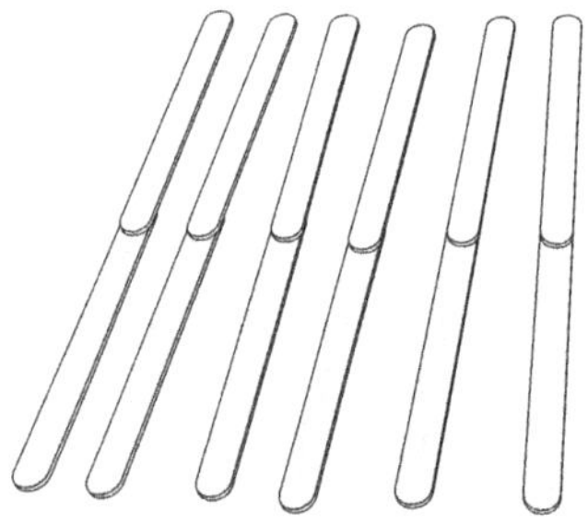

1 Using 12 popsicle sticks, make six long pieces by gluing 2 sticks together lengthwise. Allow to dry.

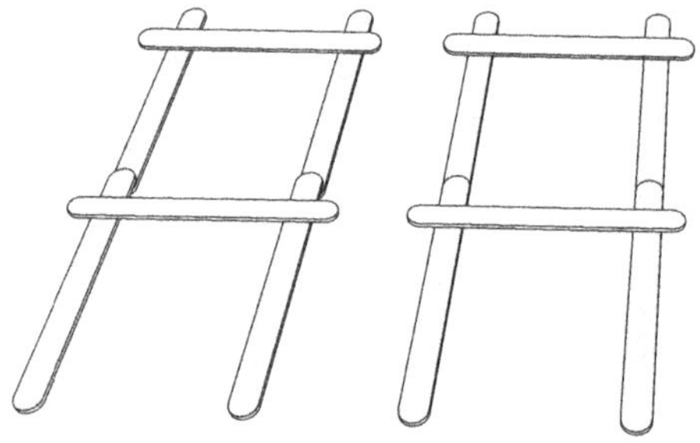

2 Glue a popsicle stick across the ends of two long pieces, leaving about 8 mm ($^{1}/_{3}$″) at the top and sticking out of the side. Next glue one across the middle. Make two of these.

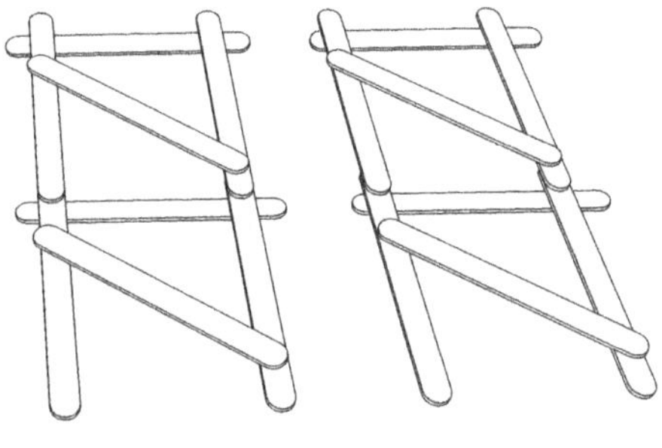

3 Glue two popsicle sticks diagonally onto each panel. Allow to dry.

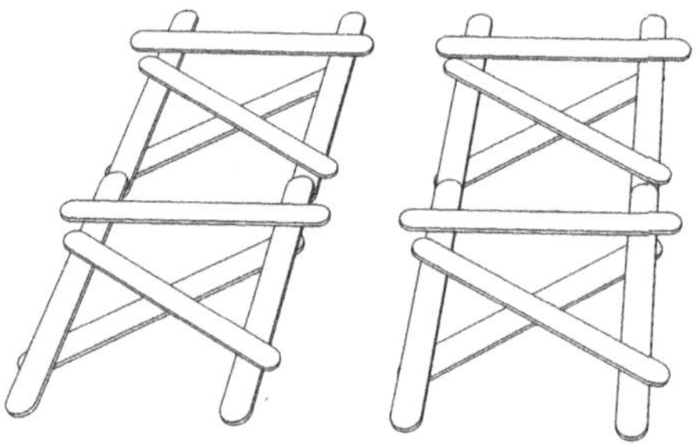

4 Turn each panel over. Glue another two popsicle sticks to cross the first ones. Do this on both panels. Allow to dry.

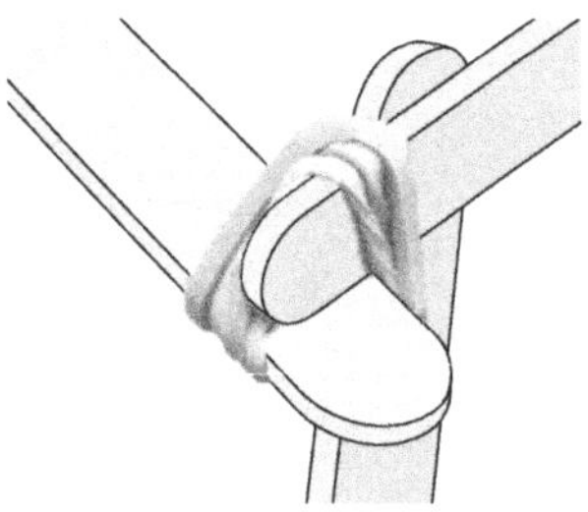

5 Attach a popsicle stick to the top of one side panel, and perpendicular to it. Wrap a rubber band around the piece that is sticking out (as shown). The new stick should be flat against the top stick and underneath it.

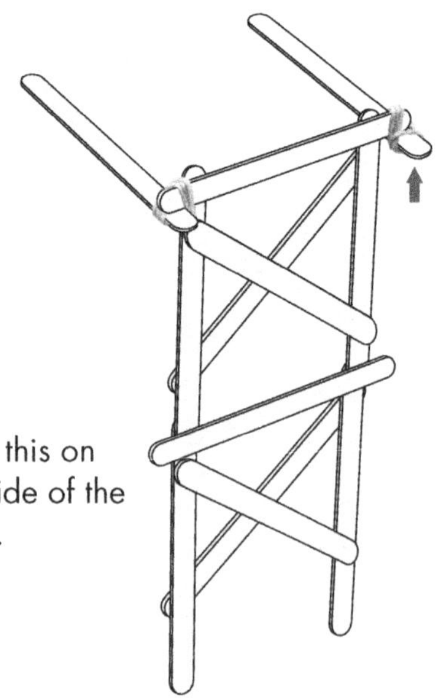

6 Repeat this on the other side of the side panel.

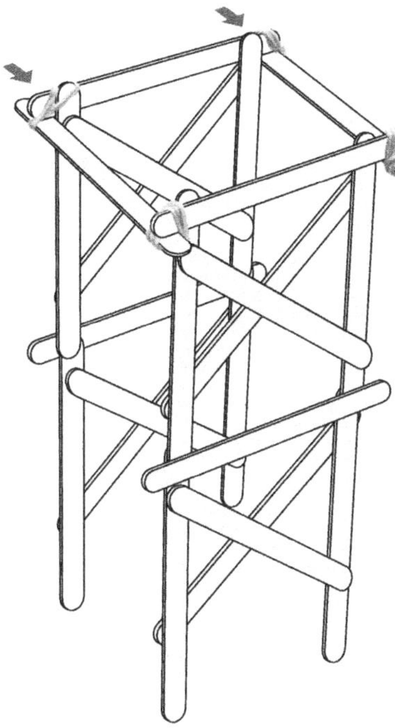

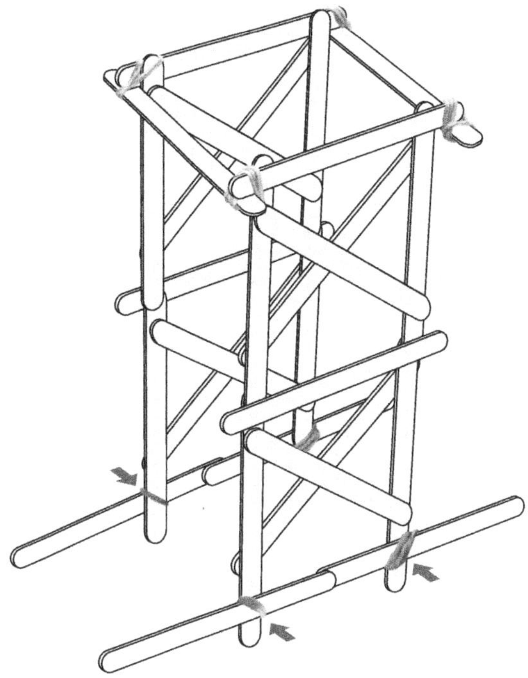

7 Secure the two cross pieces to the other side panel in the same fashion. You may need someone else to hold the frame while you wrap the rubber bands.

8 Using rubber bands, attach the last two long pieces to the bottom of each panel, so their length runs in the same direction as the panels.

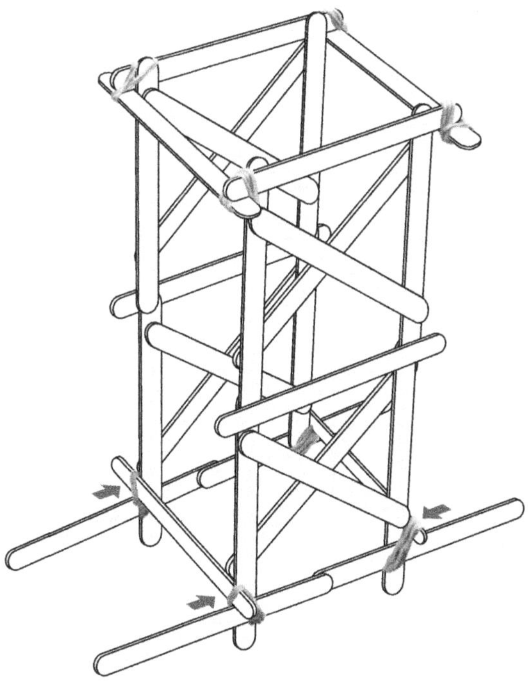

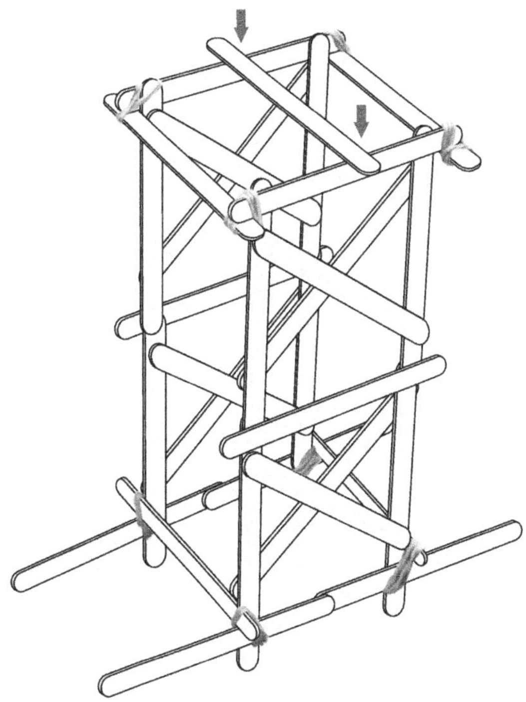

9 Attach two sticks to the "open" sides of the framework with rubber bands, i.e. perpendicular to the panels.

10 Glue one stick to the top of the framework in the middle of the other two sticks. Allow to dry.

11 Glue the fridge magnet to one end of the length of dowel. Allow to dry.

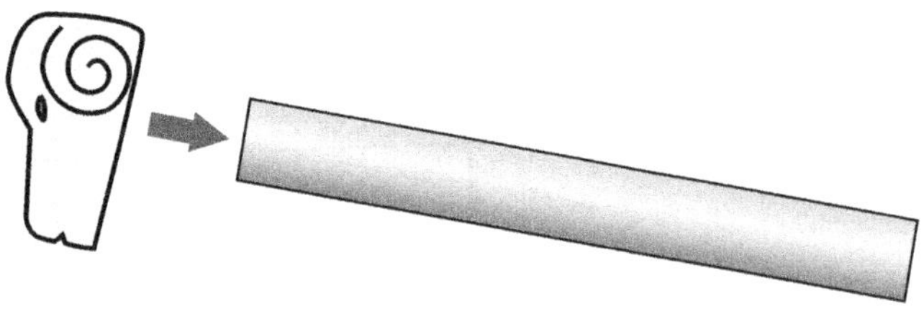

12 Tie the twine in a half hitch around the dowel.

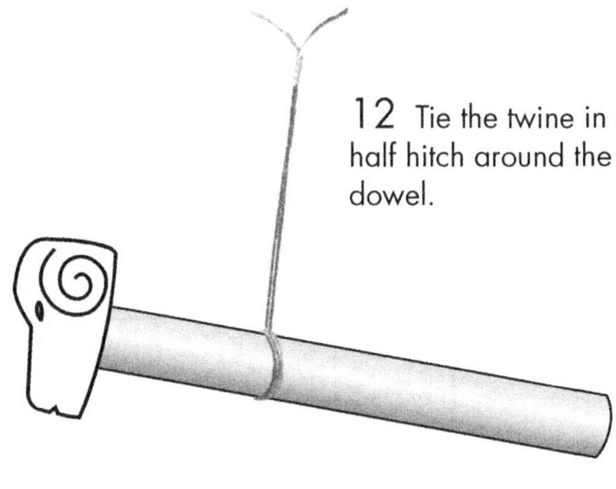

13 Tie the free end of the twine to the top center piece at a length that allows the dowel to swing when lifted at the bottom – approximately 10 cm (4″).

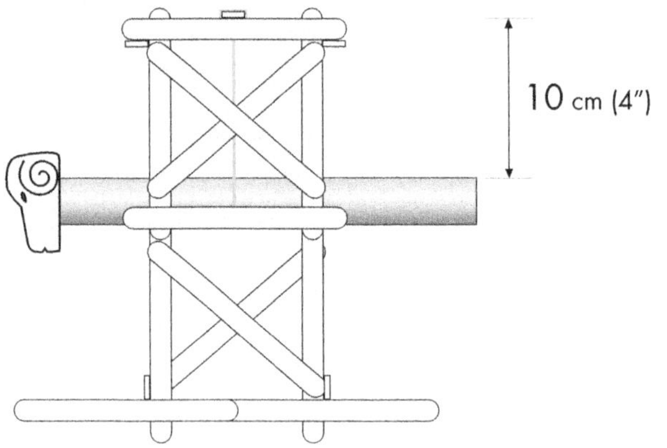

14 Neatly trim twine.

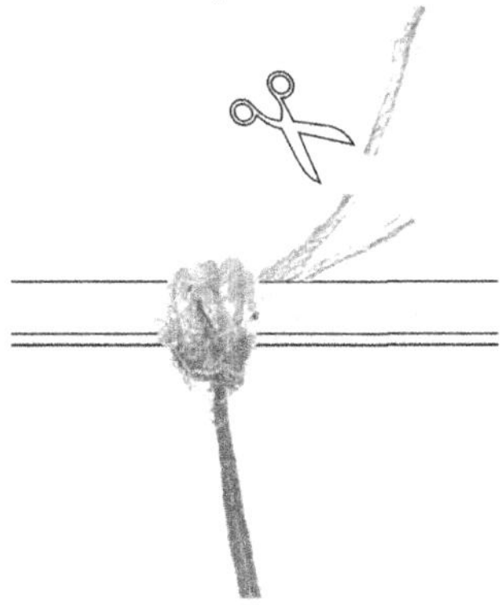

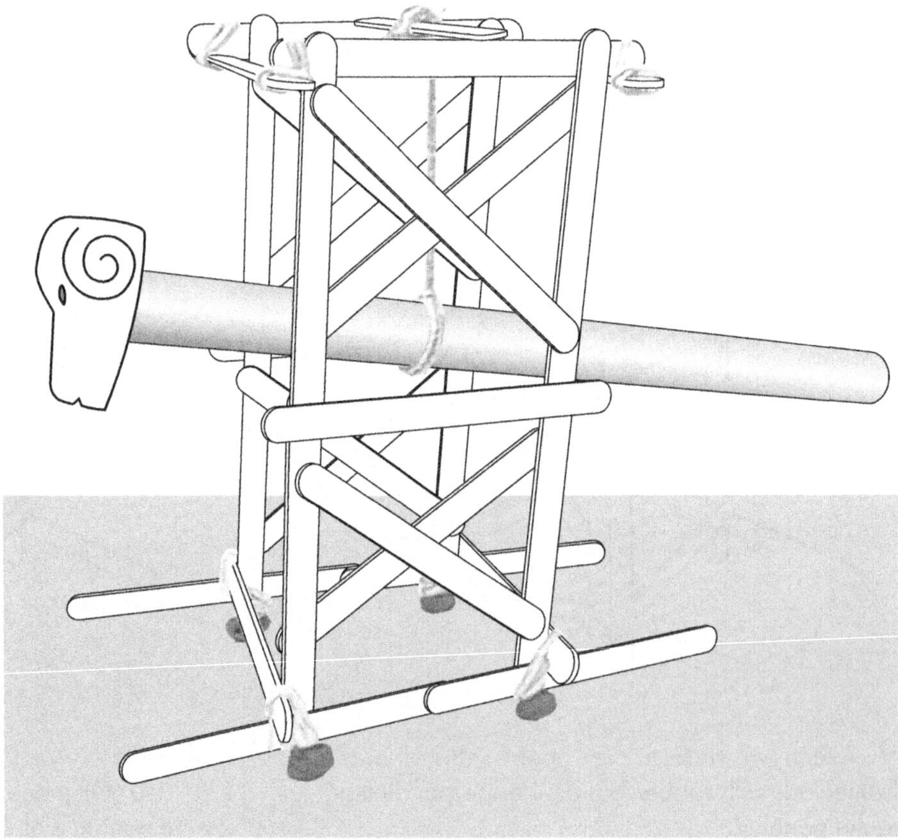

15 Divide the modeling clay into four small balls. Fix one ball to each leg of the frame. This will allow you to hold the frame still, even if the legs have turned out somewhat uneven. It will also temporarily attach the frame to the surface, so you can swing the battering ram back and forth.

16 An alternative is to do a double-rope ram. Tie two strings to the ram and tie to each top bar so that the ram is parallel to the ground.

Compare the single and double rope rams.

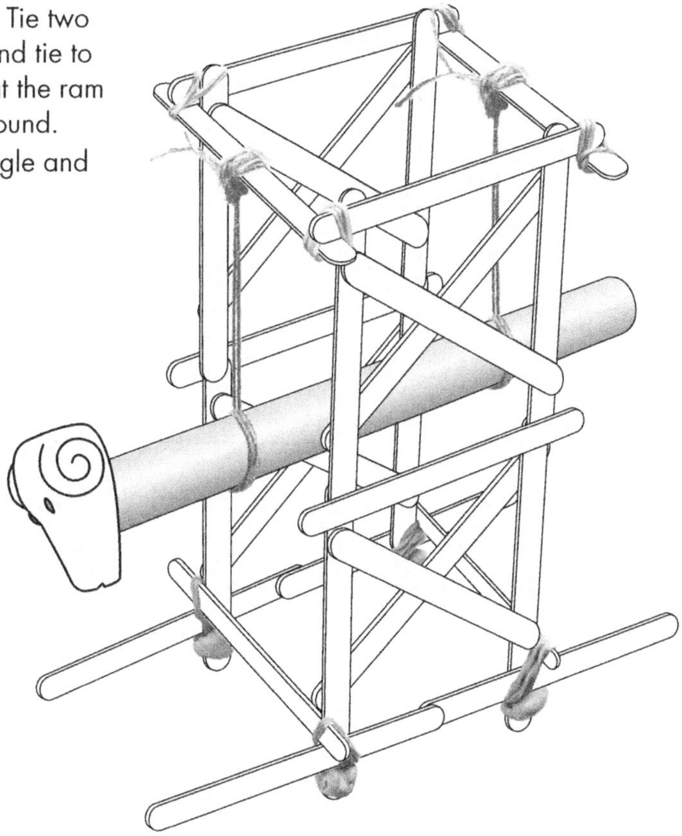

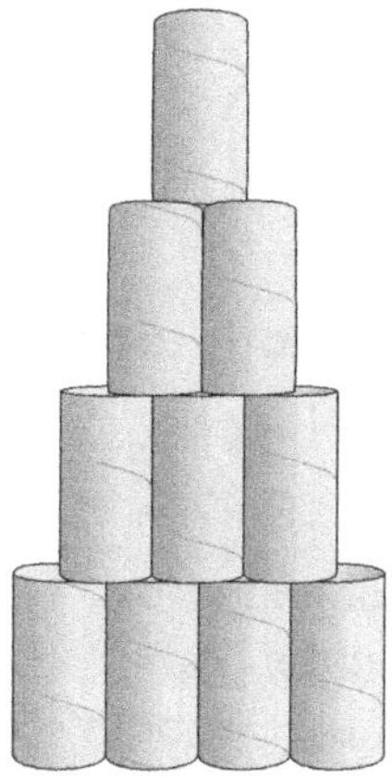

17 Build a wall out of empty toilet rolls. Move the battering ram up next to it. Use the ram to knock down the wall.

energy

Things you need

- plastic bag (e.g. shopping bag)
- lots of things to put in the bag (e.g. blocks, balls of different masses and sizes)

Words to use

faster
slower
more mass
less mass
energy

Extension

- Use objects with more mass or less mass.

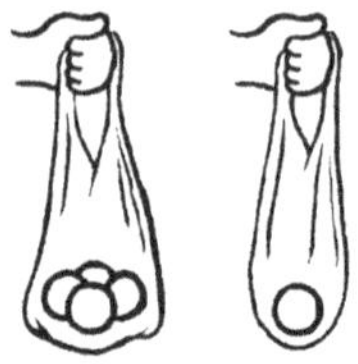

- Vary the amount of winding.

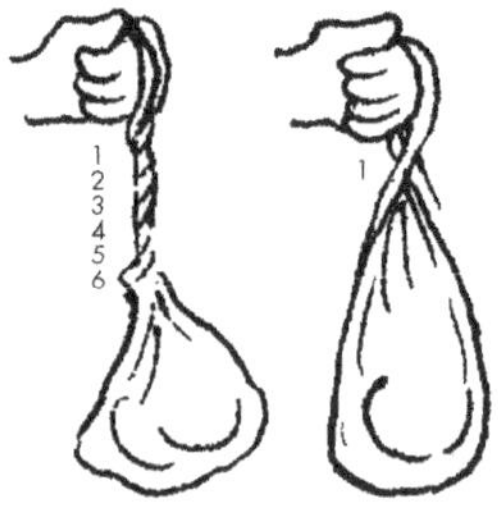

Storing energy

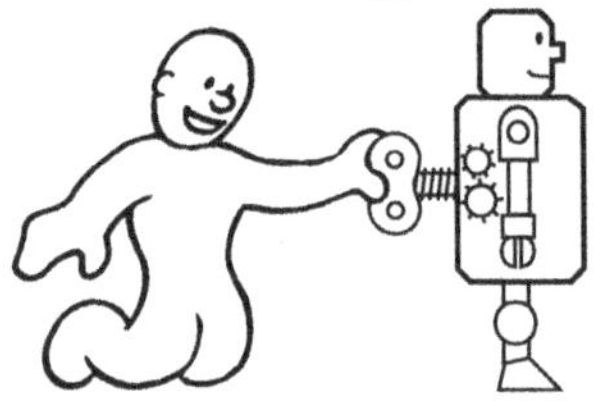

Releasing energy

Hold an empty bag by the handles with one hand.
Twist the empty bag 10 times.
Let go. What happens? ______________________

Put some mass in the bag — Twist the bag around — Let go

What happens now when you let go of the bag?

____

Which has the most energy?

____

If you were Vitruvius designing a battering ram for battle, what would you make it out of? Why?

____

Research

What forms of energy were available to the early Romans?

____

What change made water and wind energy economical?

____

storing and **releasing energy** 1

Things you need

- 2 chairs
- meter-stick
- tape
- cotton thread
- 5 or 6 washers

Words to use

moving
pendulum
weight
store
release

Extension

- Repeat the experiment, but change the length of the string. What effect does that cause? How would that principle be put to best use in your battering ram?

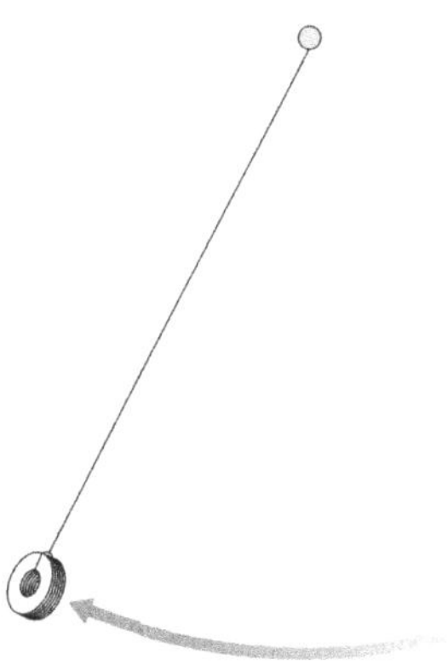

How can you store energy and release it when you choose?

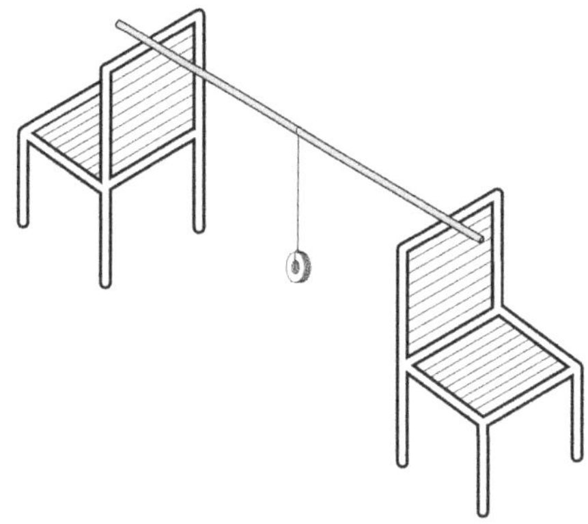

Make a pendulum.

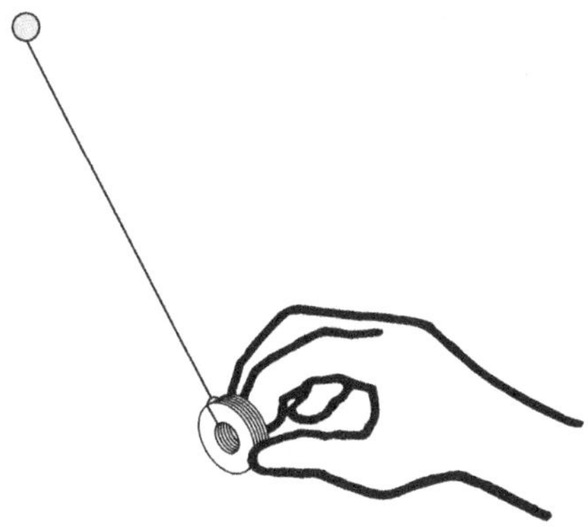

Pull back the pendulum. As you do this you are **storing** energy.

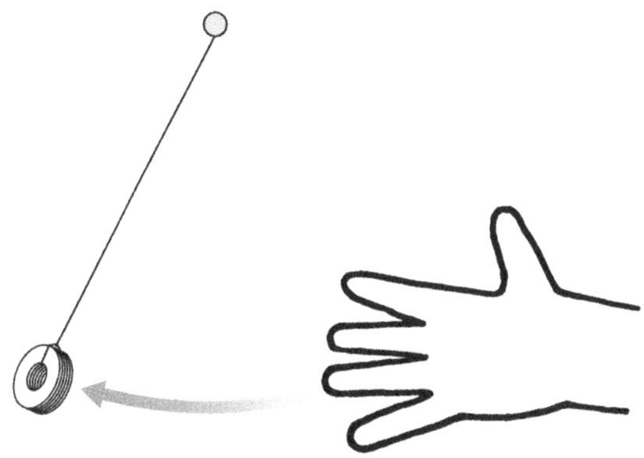

Let it go. When you do this you are **releasing** energy.

At what stage in its swing is the pendulum storing energy?

___

At what stage in its swing is the pendulum releasing energy?

___

If you were Vitruvius and wanted your battering ram to work most effectively, where in its swing would you want to place its target? Why?

___

How would you maximize the force the battering ram is able to exert on its target? Why?

___

gravity 2

Things you need

- 18 large blocks or containers, e.g. 150 ml juice cartons (full)
- sticky tape
- large hardcover book OR clipboard
- ruler

Words to use

tip
angle
measure
height
force
gravity

Extension

Objects with a height that is ten times their width are quite unstable. Why then don't people fall over all the time?

How far can you tip it before it falls?

Carefully lift up the board.

Measure the height of the board with a ruler at the point where the blocks fall off.

Tape blocks together so they don't slide off each other.

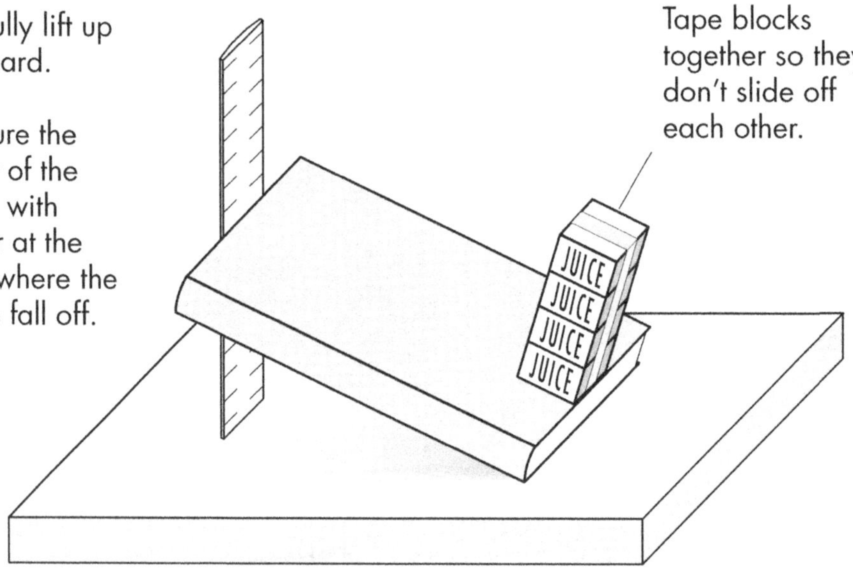

Do three trials of this and record your results below.
Add two blocks to the stack and repeat the experiment.
Try it again with two more.

	Tower 4 high	Tower 6 high	Tower 8 high
Trial 1	_______ cm	_______ cm	_______ cm
Trial 2	_______ cm	_______ cm	_______ cm
Trial 3	_______ cm	_______ cm	_______ cm
Average	_______ cm	_______ cm	_______ cm

Explain your results: ______________________________

____

If you were Vitruvius knocking down a fortress with a battering ram, how would this information be helpful to you? Why?

____

For information on Vitruvius see back cover.

section six

onager

project 7 **onager**

Materials

- 20 popsicle sticks
- 1.3 m (4′) cotton string
- craft glue
- 15 mm binder clip
- shears or scissors (for an ADULT to use)
- clothespin (remove and use the spring wire)
- paperclip, cut to 10 cm (4″)
- small needlenose pliers
- plastic ziplock sandwich bag

1 Cut 7 popsicle sticks cut in half (ask an adult to cut them with a shears or scissors or strong scissors). Cut one of those halves to 40 mm (1 1/2″).

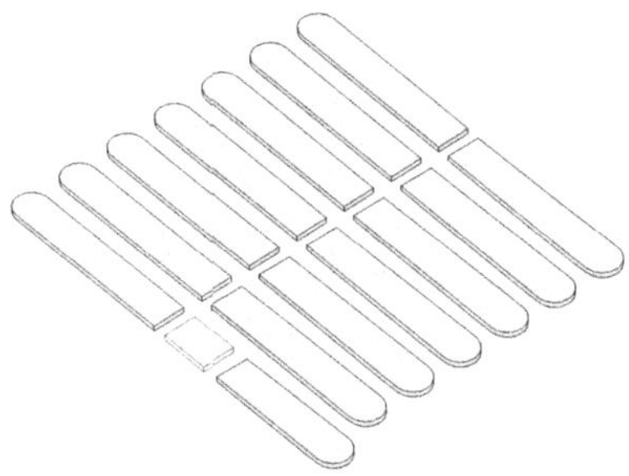

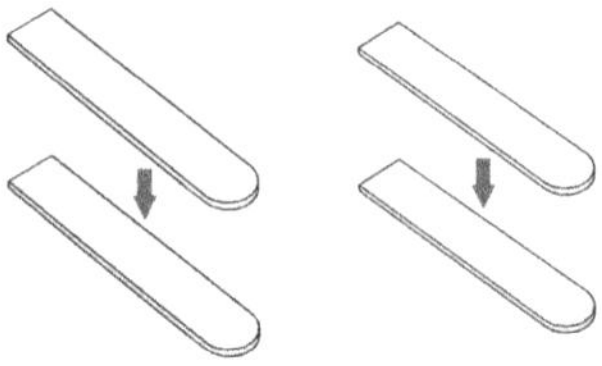

2 Glue two half sticks so you have one piece, two sticks high. Make two of these.

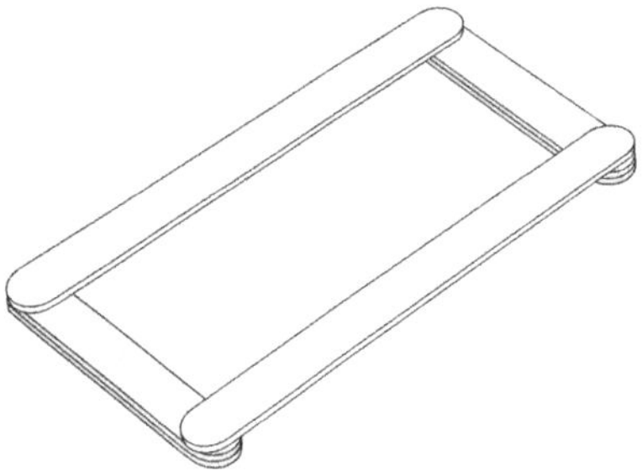

3 Join together with 2 long sticks.

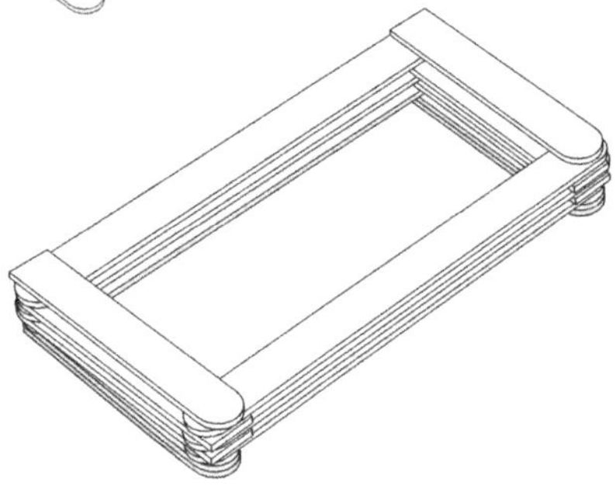

4 Add alternate layers of half sticks and full sticks (as shown).

5 Glue two half sticks on the inside of the rectangle, at one end. Be sure the square end is facing down. These are the uprights. Allow to dry.

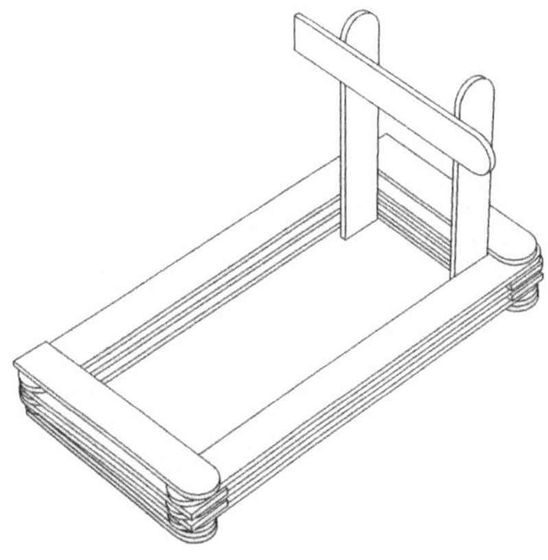

6 Glue a half stick across the end of the uprights. This is the crosspiece.

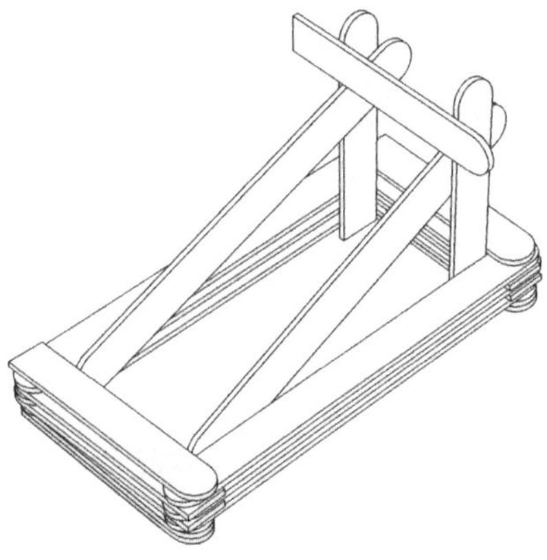

7 Support the uprights with two diagonal sticks under the crosspiece. Allow to dry.

8 Fold a string in half 4 times. You will have 8 parallel lengths.

9 Put your index finger through the loops on the right hand side (as shown above). Hold the other ends in the other hand. Twist by twirling your index finger until the twists are along the whole length of the string.

10 Wrap the twisted string around the middle of your onager then tie together with a knot at the side.

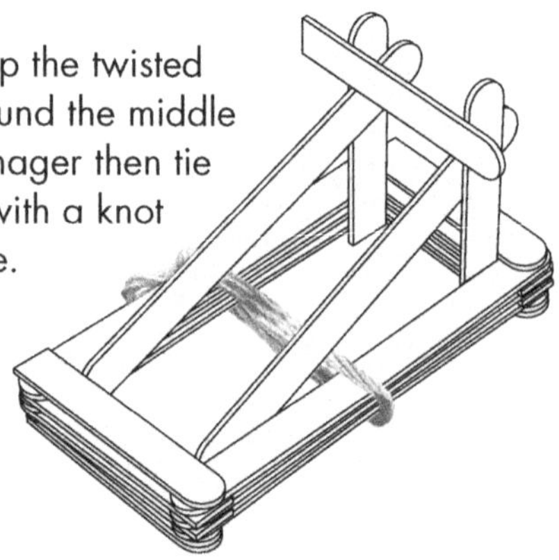

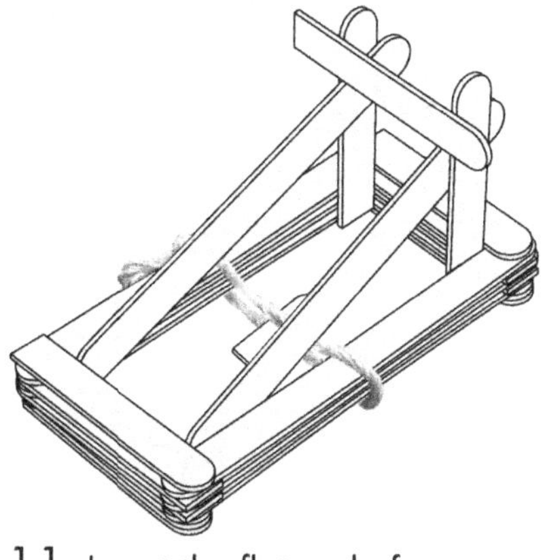

11 Insert the flat end of the quarter stick between the string. This is your twister.

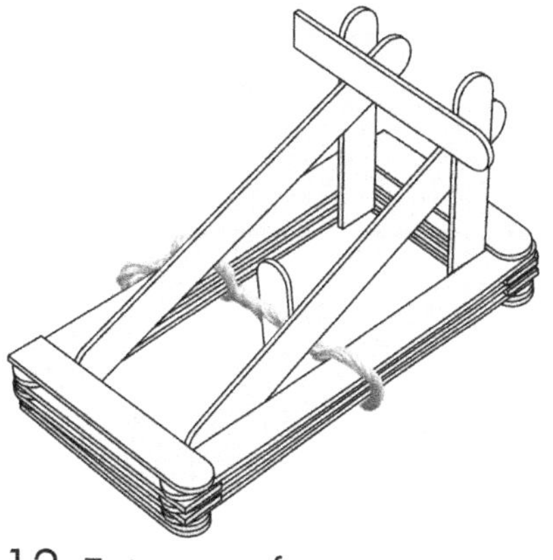

12 Twist away from the crosspiece until it is very tight.

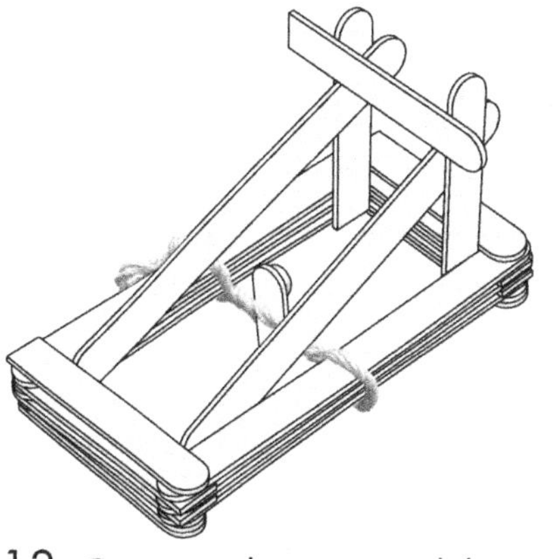

13 Once you have twisted the string, put a large drop of glue on the top side of the twister, on the side facing the crosspiece.

14 Place the remaining whole stick against the glue. Hold in place for a minute or so, and then secure in place with the binder clip. Allow to dry well. The arm will hit against the crosspiece when you let go.

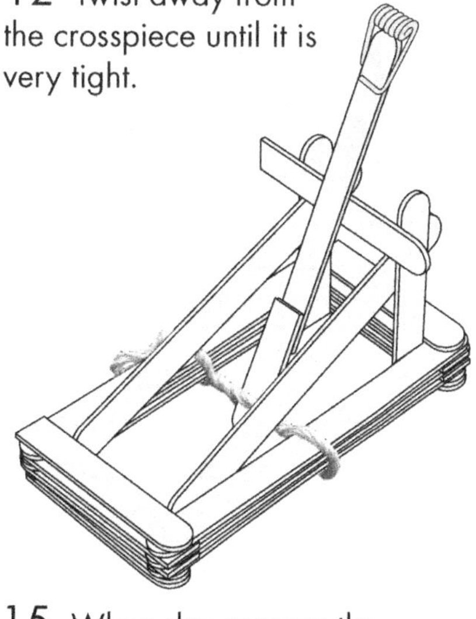

15 When dry, remove the binder clip. Attach the coil of wire (from the peg) to the top of the stick. Add some glue to secure.

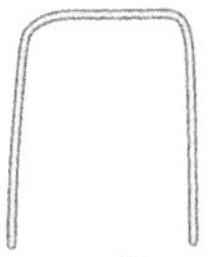

16 Straighten out a paper clip with pliers. Then bend it to a handle shape.

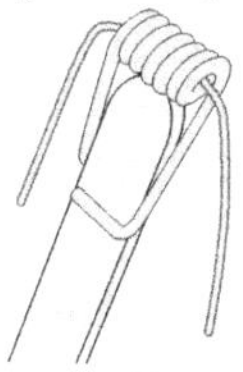

17 Thread through the clothespin coil. You may have to straighten one side to get it through and then re-bend it.

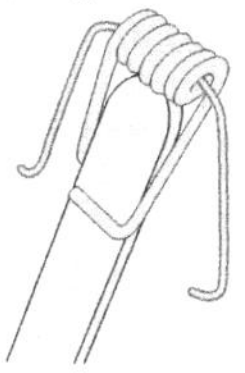

18 Curl the ends.

19 Cut out a piece of plastic. Use the thicker plastic near the zip lock of the sandwich bag.

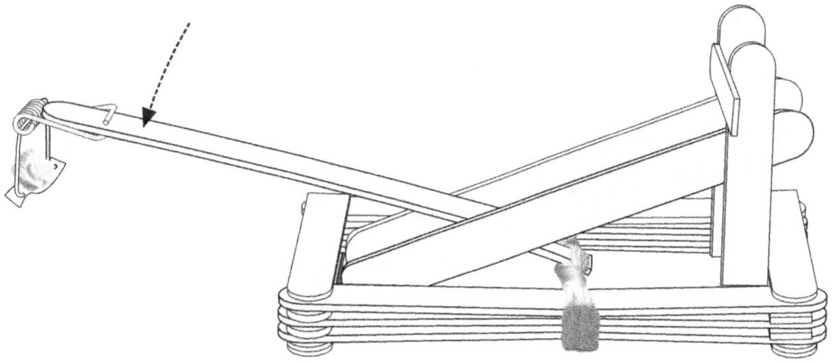

20 Poke holes in the plastic then hook it on to the wire.

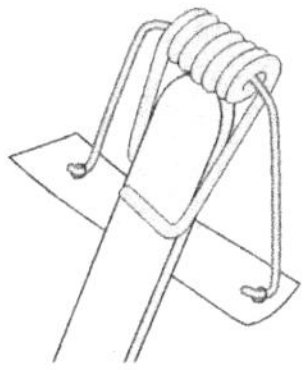

21 Make a small paper ball and put it in the cradle. Pull the stick back (don't touch the clip) whilst holding the base in your other hand.

22 To shoot, let go.

Warning: Use only under adult supervision. Do not fire heavy objects. Never aim at a person.

Extension: Build the frame of your onager in the same manner as before. Instead of adding the sling and coil on the end, glue a mini-cupcake paper liner onto the end of the popsicle stick. Put a ping pong ball in the cupcake paper and shoot at a target. This type of catapult is known as a *mangonel*. (IMPORTANT: Never aim at a person!) Note: if the cupcake paper is thin, use two stacked together to make it more stiff.

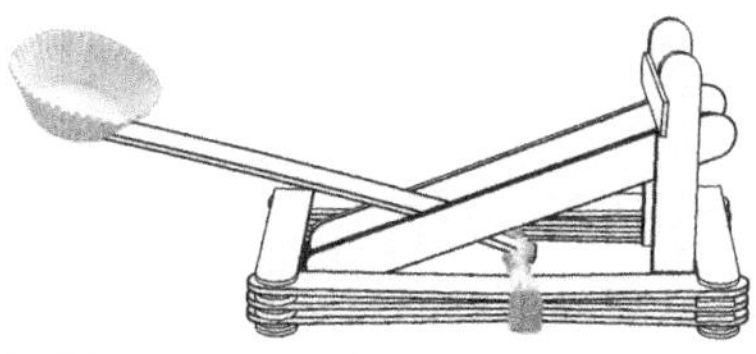

the onager

the scorpio

THE ONAGER

The onager is named after a draft animal, the Asian wild ass, well known for its very powerful kick back. This war machine used towards the end of the Roman Empire was the most powerful instrument of war used in sieges.

From technical testimonies we learn that nine people were needed to operate just one of these machines. The projectiles ranged from a weight of 5 kg to 50 kg (11 to 110 lb), so the shooting range varied from 200 to 600 m (219 to 656 yards) according to the weight of the projectile. The 'kick' of the machine was so powerful that if it was not well harpooned to the ground it would jump backwards after each shot, risking tipping over, not to mention harming the operator.

THE SCORPIO

A ballista is a two-armed torsion catapult capable of firing bolts or stones with considerable force and accuracy. These were built in various sizes and most often used in sieges.

The scorpio was the light bolt-shooting ballista and was more of a sniper weapon than a siege weapon, being operated by two men only, although it was also used in sieges. A weapon of great power and precision it was particularly feared by the enemies of the Roman Empire.
It was the forerunner of the crossbow, probably inspired by the Greek *gastraphetes*, a small portable ballista powered by darts. It was invented in 50 BCE, with darts 70 cm (27″) long and a throw precision of 100 m (110 yards), but could reach with less precision up to 400 m (437 yards). Alternatively they could use darts of 130 cm (51″) with a cast of up to 650 m (710 yards), but with less shooting precision.

During the Republic and the Empire of the early centuries, each legion had 60 scorpios which could fire up to 240 arrows per minute. The scorpio had two main functions: a flat trajectory which could bring down a man at 100 m (110 yards), piercing even the shield, or a parabolic throw, with distances up to 400 m (437 yards).

storing and **releasing energy** 2

Things you need

- ballpoint pen, spring loaded with little tab on the side that releases it
- small piece of paper rolled into a ball

Words to use

kinetic energy
potential energy
elastic potential energy
stored
released

How can you store energy and release it when you choose?

Use a "click pen" to shoot a paper ball.

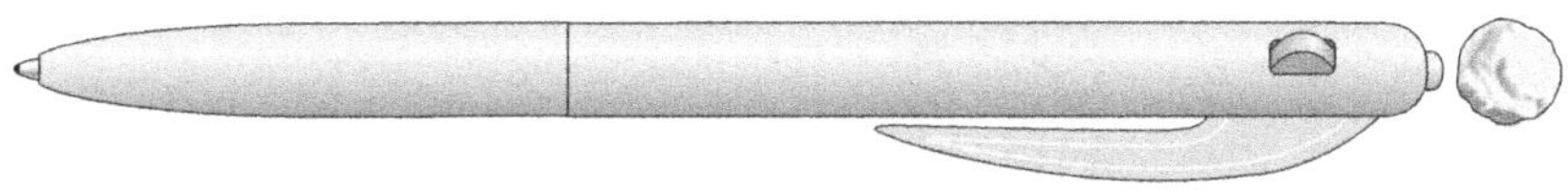

When you store energy it is called **potential energy**.

When the potential energy is converted into movement it is called **kinetic energy**.

Before you press the "tab", what sort of energy does your "click pen" have?

____

After you press the tab, what sort of energy does it have?

____

What kind of energy does your onager have when you pull back the arm?

____

What kind of energy is there when you let go?

____

gravity 3

Things you need

- rubber bands, size 64
- 15 cm (6″) ruler

Words to use

angle
distance
fall
gravity
force
mass
launch
elevation

How long will it take to fall?

Make a rubber band launcher.

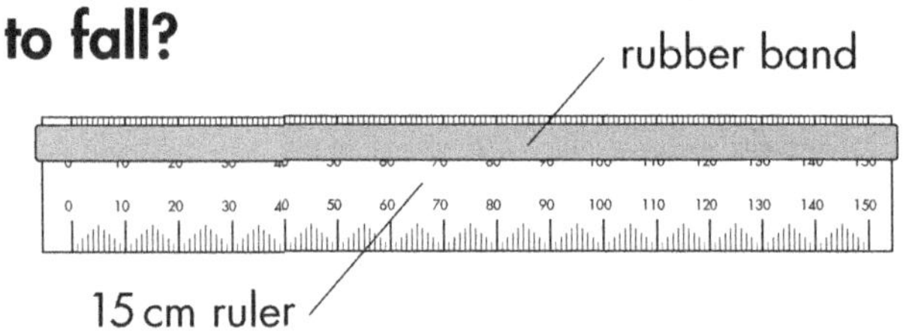

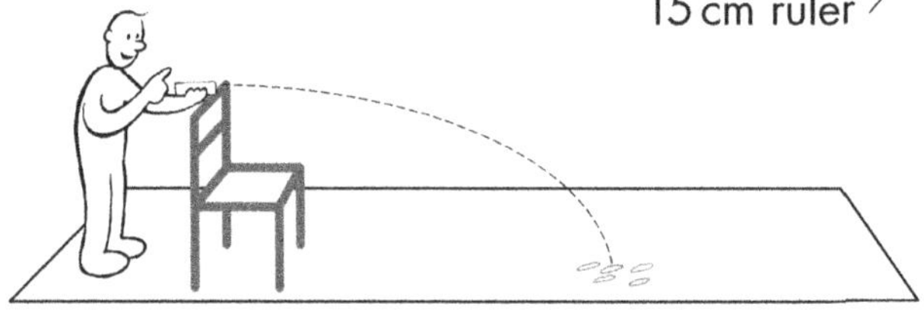

Hold the launcher level and practice shooting rubber bands until you can shoot consistently.

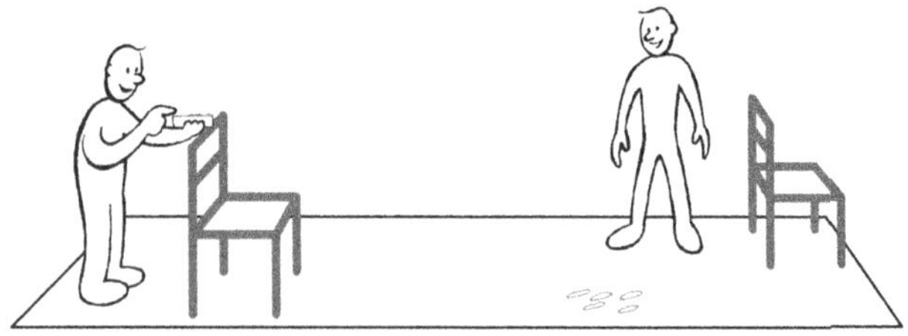

Have someone stand at that distance, but well to the side so they don't get shot with a rubber band.

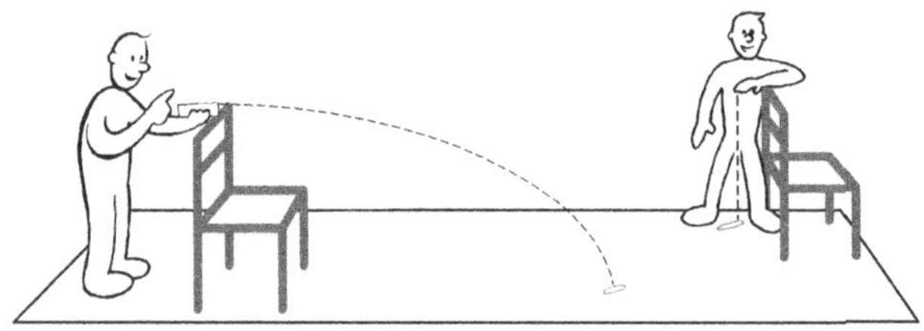

At the same time, the two people shoot the rubber band and drop the rubber band. (Both people drop from the same height by supporting their hands on chairs of equal height.)

What happens? ______________________________

____

How can you explain this? ______________________________

____

Your onager has similarities and differences to your rubber band launcher. List three similarities and differences.

____

____

____

levers 1

Things you need

- tongs
- see saw
- garlic press
- nut-cracker
- scissors
- stapler
- scales

Words to use

lever	class
fulcrum	type
load	
effort	

Extension

Make a list of other levers you have seen, what they are used for, and what is their class of lever.

There are 3 basic kinds of levers

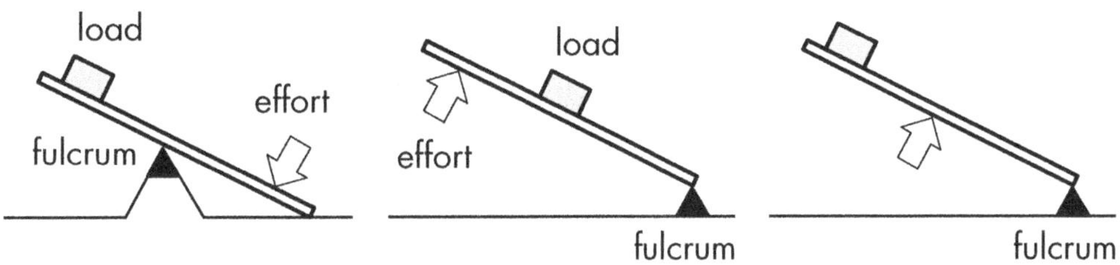

class 1 lever **class 2 lever** **class 3 lever**

Which class of lever are each of these things?

tongs

Class ___

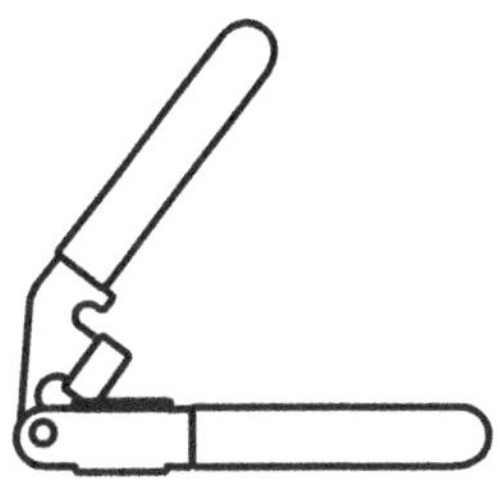

garlic press

Class ___

scales

Class ___

scissors

Class ___

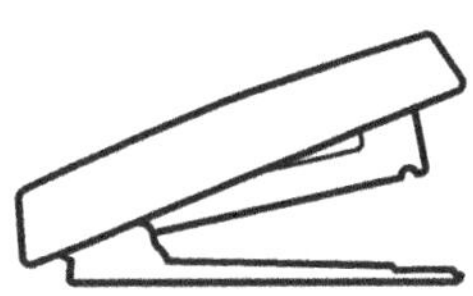

stapler

Class ___

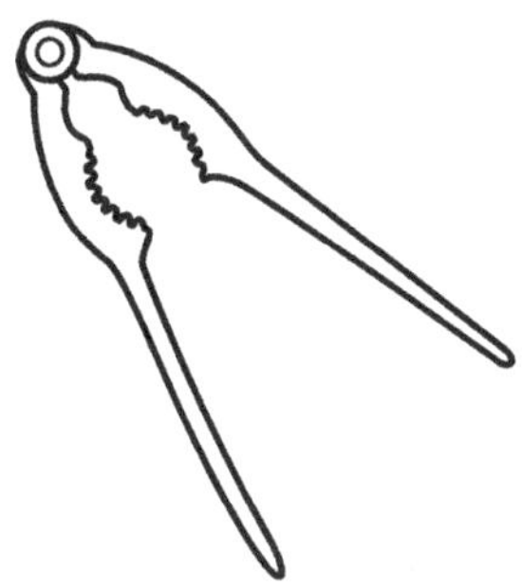

nut-cracker

Class ___

levers 2

Things you need

- scissors
- scrap paper
- nut cracker
- nuts in shell (e.g. walnuts)
- tongs
- marbles
- 2 plastic containers

Words to use

fulcrum
load
effort

Extension

Have a party.

Use different types of levers for preparing and serving the food. List the utensils you use below.

____

____

____

____

____

"Give me a firm spot to stand on and I will move the earth."

– Archimedes
c.220BCE

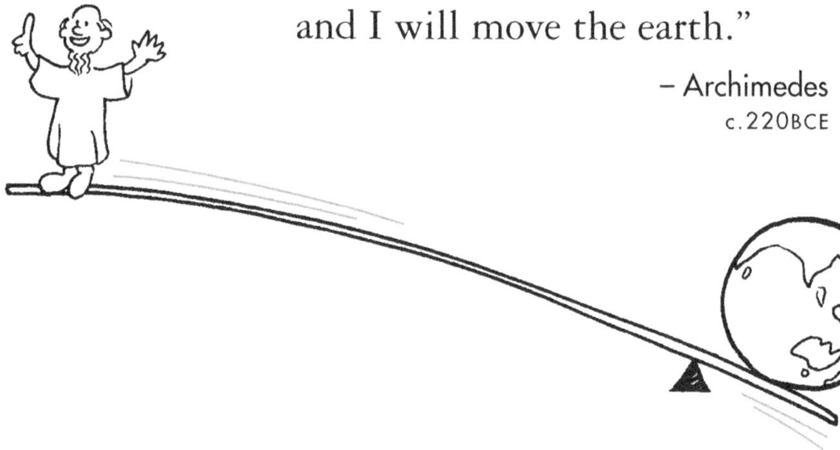

Set up three stations with the three types of levers. Try each.

Class 1
Cut out a shape from a piece of paper using scissors.

Class 2
Crack open some nuts.

Class 3
Use tongs to transfer marbles from one container to another.

Which type of lever is your onager? ________________________

section seven

rolling tower

project 8 rolling tower

Materials

- 2 egg cartons
- 70 popsicle sticks
- 86 craft matchsticks (headless)
- string, 15 cm (6")
- craft glue
- 5 pencils, circular in cross section
- gaffer tape OR duct tape
- scissors
- ruler

Back panel

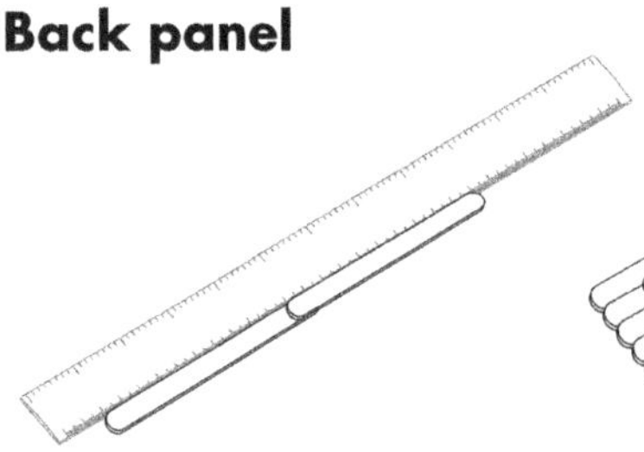

1 Overlap two popsicle sticks and glue them together. Line up against a ruler to keep them in a straight line. (Don't glue it to the ruler.)

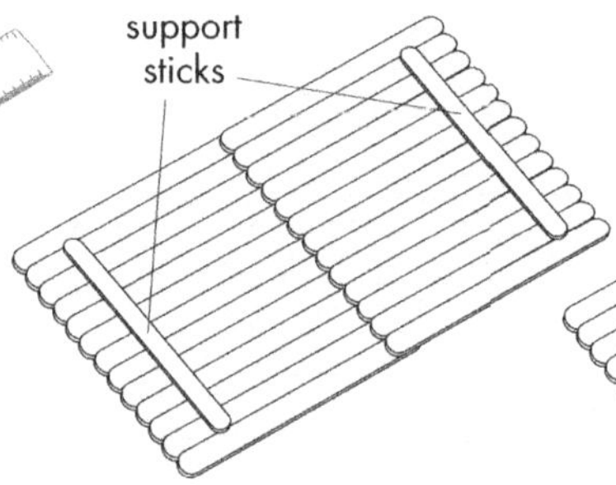

2 Glue more sticks against them in a line until you have 13 pairs. Glue on 2 sticks as support. This is the back panel.

Top & bottom panels

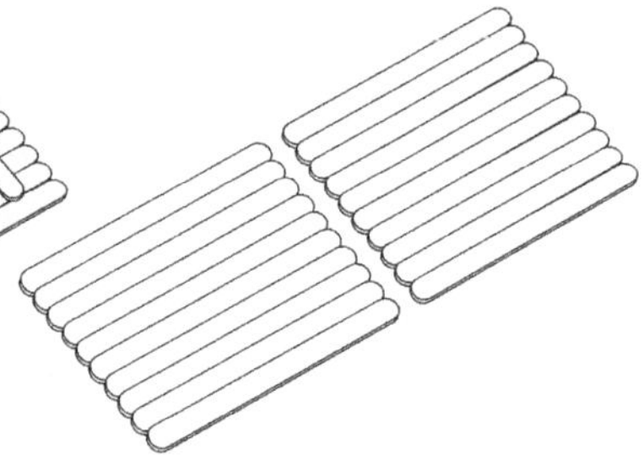

3 Glue two sets of 10 sticks together side to side. These are the top and bottom panels.

Side panels

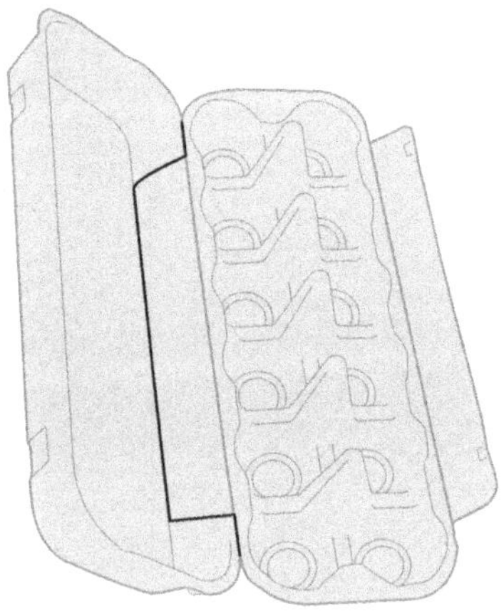

4 Draw a line on the lid of 2 egg cartons as shown.

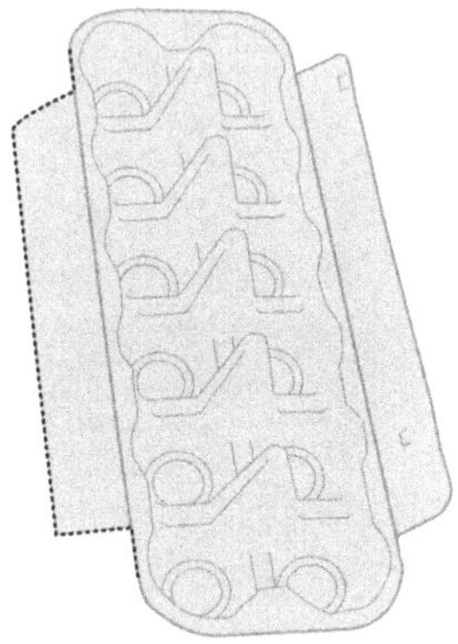

5 Cut the lids off on this line.

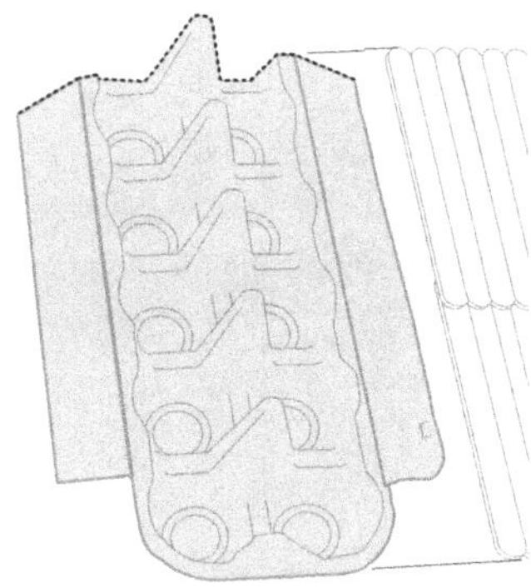

6 Line up the back panel and cut off the carton to this length.

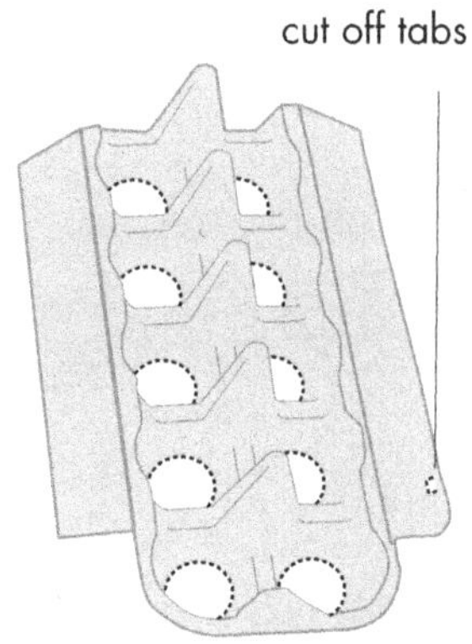

7 Cut the bottoms off each egg cup. Also cut off the tabs on the flap so this side is smooth.

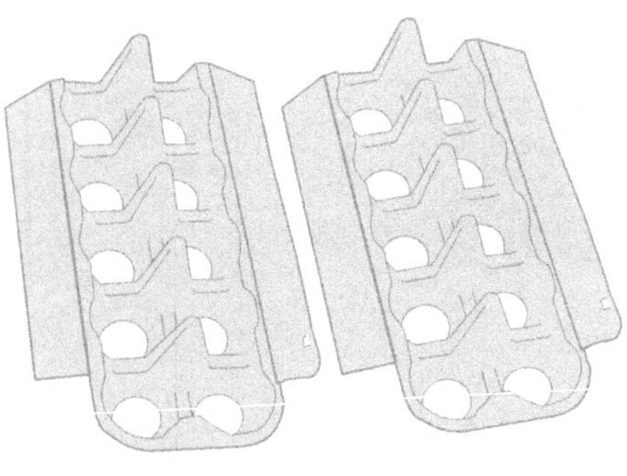

8 Make a copy of this carton. You now have made two side panels.

Front doors

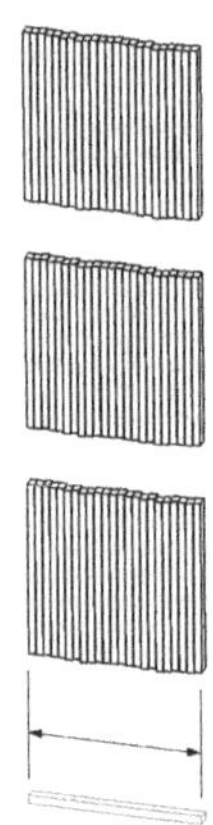

9 Glue 3 sets of approx. 22 matchsticks together to make the length of a matchstick.

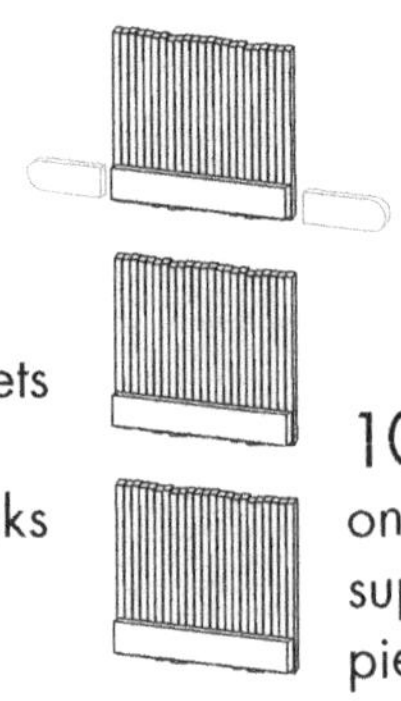

10 Glue on a supporting piece of popsicle stick.

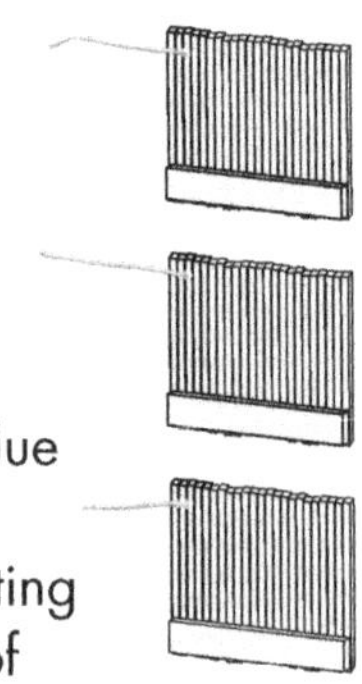

11 Cut 3 lengths of string, each approx 25 mm (1"), and glue to each door.

Front panel

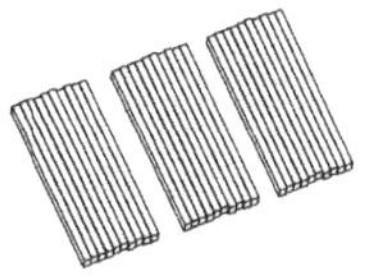

12 Glue 3 sets of 10 matchsticks together. These are the connecting pieces.

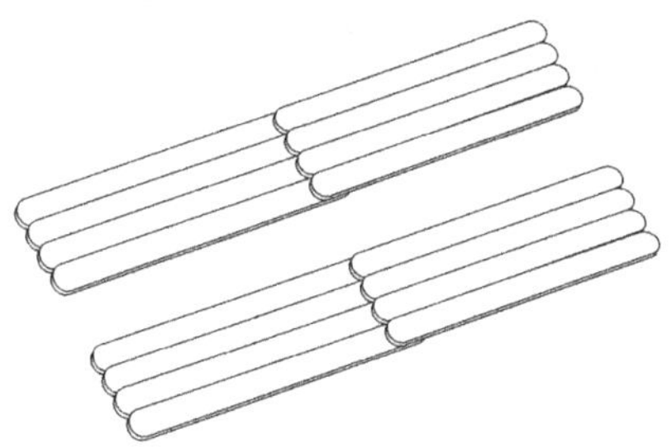

13 Glue two sets of 4 pairs of sticks together side by side.

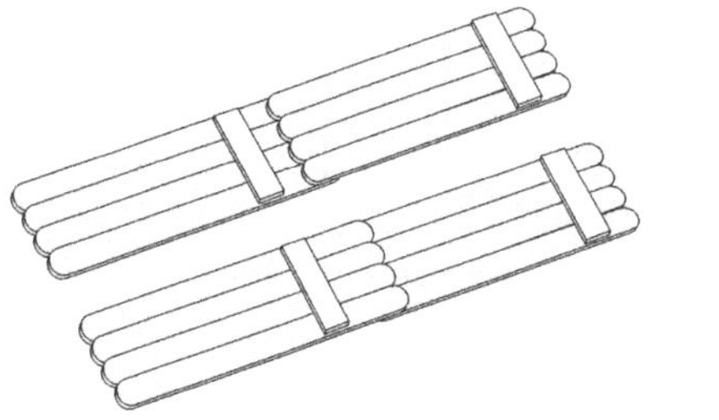

14 Cut some popsicle sticks to the width of 4 sticks. Glue them on as support pieces. Make sure they do not overlap the edges. These are the front uprights.

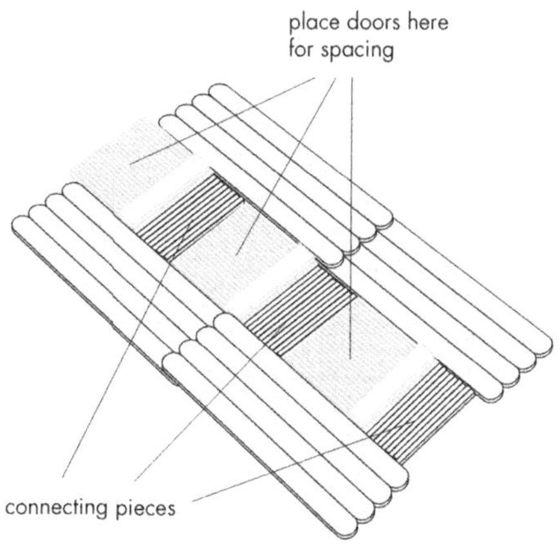

15 Glue the three front connecting pieces to the two front uprights. Place the doors in position to help with spacing the connecting pieces. (Don't glue the doors).

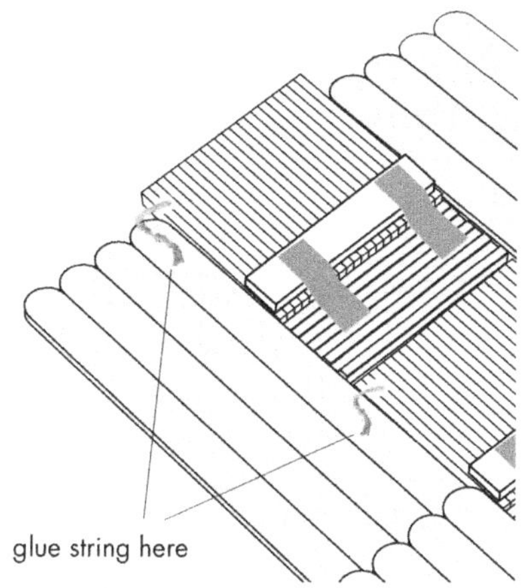

16 Tape the doors onto the connecting pieces with gaffer tape. Glue the other end of the string to the side panel. You now have a front panel.

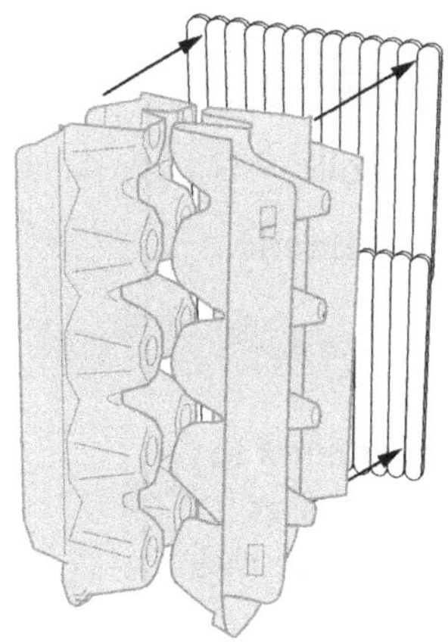

17 Glue the back panel to the flaps of the egg cartons. Make sure they are aligned and don't stick out beneath the bottom of the back panel.

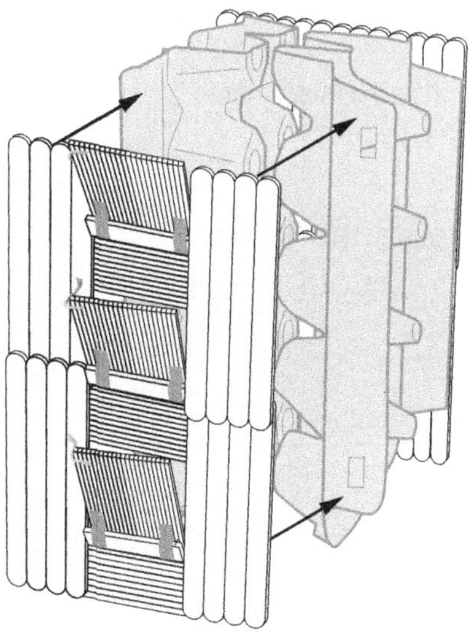

18 Glue the front panel to the egg cartons.

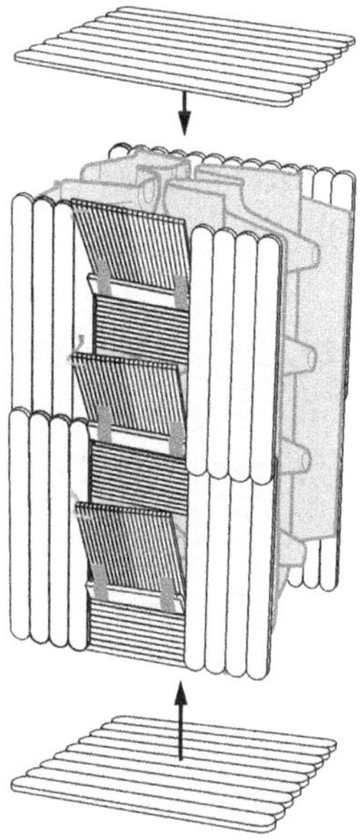

19 Glue the bottom panel to the base of the egg cartons. Glue the top panel on top.

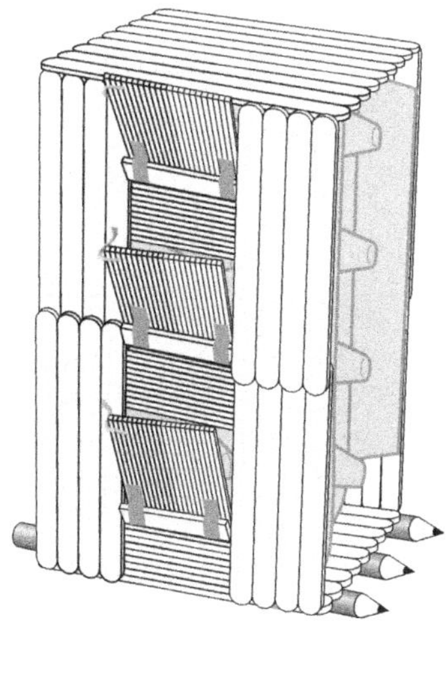

20 Use pencils underneath as rollers.

wheels

Things you need

- 1 block of wood approx. 8 cm x 10 cm x 20 cm (2″ x 4″ x 8″)
- 4 equal sized metal lids
- 8 washers
- 4 short nails
- 4 rubber bands
- ruler
- hammer (be careful not to hit your fingers when hammering)

Words to use

roll	faster
rubbing	slower
friction	energy
gripping	axle
turn	

Build a cart

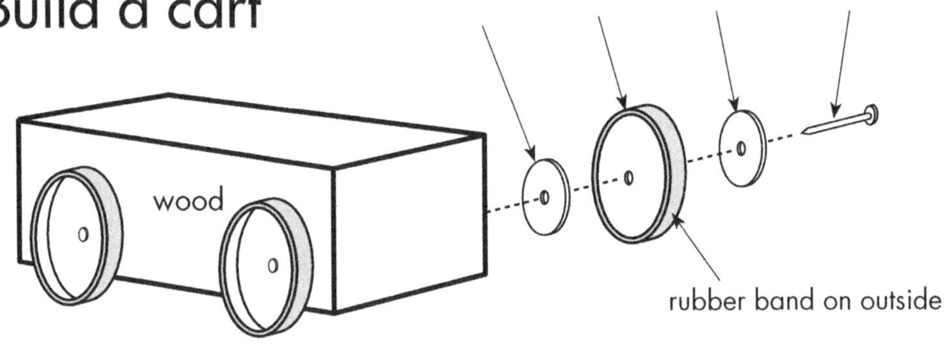

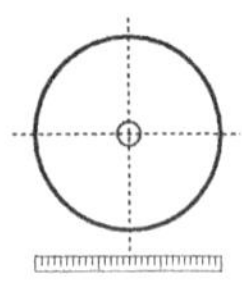

Work out the center by finding the diameter. Punch a hole right in the middle of the lid.

Don't pound the head of the nail into the lid. The lid must be able to rotate freely.

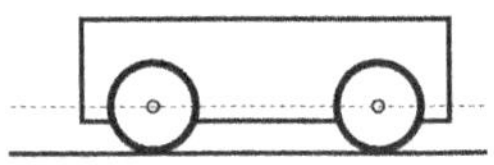

Make sure the nails (axles) are all even and that the wheels all sit flat.

The nail acts as an **axle** to let the wheel turn.

The wooden block is the **load** which is moved by the wheels.

Push the cart along the floor or roll it down a ramp.
If the wheels don't turn easily, make adjustments so they do.

When the Romans built a rolling tower, they used logs underneath the base to roll the tower.

In your model, you used pencils underneath the tower.
What was the shape of the cross section of your pencils?
Why was that important?

____

____

Replace those pencils with some that are hexagonal in cross-section. Compare your ability to roll the tower now, to rolling it on the first pencils.

teacher's notes

Some suggestions for good model-making:

Note: Work out a strategy for preparing the pieces and plan ahead to suit your schedule.

1. Always make the model first yourself if you are going to have a class make it.
2. Be as precise as possible. Pieces have to fit together, so it is important to follow instructions carefully.
3. Make careful measurements.
4. Be sure to allow time for glue to dry. In some cases, you can proceed to the next step if the piece is only a bit dry. Other times you must allow it to dry overnight.
5. Plan your steps in advance so you can have pieces ready when you need them.
6. To glue two sticks together in a straight line, place them up against a ruler.
7. Some popsicle sticks can be irregularly shaped, or missing a bit out of them, or are warped. Some irregularity is ok, but check through your sticks to make sure you have removed the ones you cannot use.

   Save the pieces, as you may be able to cut them up and use them when small pieces are needed.
8. Gather a collection of weights (rocks, small tins of food, hammer) which can be used for holding models while glue is drying.
9. Gluing technique: put glue on the stick, place second stick carefully, then hold the sticks together for 30 seconds to a minute before you let go. Then place appropriate sized weights on the sticks. You may also need larger objects on which to lean the drying model, e.g. the pile driver.
10. To cut egg cartons, use short, pointy scissors. Get an adult to do this if necessary.
11. Get an adult to use the shears or scissors to cut the popsicle sticks.
12. To glue the tiny ropes on the Rolling Tower doors, hold the end of the string on the spot of glue with a matchstick. Then, place an extra drop of glue on top, once it has set a bit.
13. Keep ends flush, unless the instructions say to poke the stick out over the edge "x" mm. Protruding sticks can mean that something will not fit later on in the project.
14. For best results, use a small pliers if you need to bend wires e.g. paperclips.
15. Pay attention to your technique. The results will speak for themselves and the projects will be real works of art!

Project 1 **abacus**

A tool which the Romans adapted for their use was the abacus. Their children, boys, used it to learn simple maths. And Plato suggested that people use it to carry out "As much as is necessary for the purposes of war and household management and the work of government." In this lesson the students will construct a simple abacus, of Roman style. The Roman abacus was made with slots, as opposed to the Japanese abacus, which was made of beads on wires. Students will learn the basic method of its use. The abacus would make a good research topic (for students wanting more of a challenge) as it has a practical component. Students can have fun getting good at using it. Draw up a worksheet of more problems to add and subtract. Include more difficult problems.

You will find that hardware stores usually have a bin full of offcuts of wood which they will give to you for free. These can be used for backing boards. It is important, however, that you check that the boards are straight, and not warped.

Project 2 **wax tablet**

Before making your model, you will have to determine whether or not your budget allows you to use beeswax. It is considerably more expensive, but check craft stores for the most economical way to purchase it. Beeswax is certainly more authentic, and is much nicer to use than plasticine. However, plasticine works well, and is a much cheaper option.

Project 3 **Roman roads**

In this project, you will be constructing a model which is similar to a section of a Roman road. The materials, while on a smaller scale, will be laid down in a way which duplicates the principles by which the Romans

teacher's notes (cont'd)

built their roads. The passage of time has shown that this style of construction has produced strong, durable roads which can still be seen today. Many hundreds of miles of roads were built in Roman times, some still surviving, and all of which "led to Rome".

Rocks, sand and gravel can all be purchased cheaply from a garden or landscape supply outlet, or depending on where you live, you may be able to find these materials outside. (Note: do not "borrow" decorative stones from your neighbour's garden!)

mass

This lesson investigates some basic properties of various rocks and solid objects. The types of objects which would be best suited for roads are the ones which are very dense or heavy. In order for the students to appreciate that not all rocks are very dense, include some samples such as pumice, which, if used to build a road, would crumble under the weight of heavy vehicles.

For the extension, you may like to refer to Mohs Scale of Hardness if you have older students.

space

In this lesson, we look at the concept of space. Generally, our attention goes to the physical objects in a space. However, there are always smaller and smaller bits into which a space can be broken up. We draw students' attention to the use of the whole of a space. The borders, or barriers of a space (the shape of the poster board shapes) define the area of the space here, then there is the mass in the space (cans or blocks) and then there is the space in between the masses. (Note that in Japanese flower arranging, the spaces in between the flowers are included as an important part of the arrangement.)

In the building of the Roman roads, a defined space (the trench) contained the large rocks and the smaller gravel and sand which filled the smaller spaces, and contributed to the strength of the Roman roads.

volume

This experiment begins to tie together the principles used in building the Roman roads. The students work with mass to understand first hand the meaning of volume. Then, they begin to make some simple measurements. Finally, they take a measurement of the volume of mass they used to make their model. They will use this figure, which they determined themselves, to ultimately calculate the density of their sample Roman road.

gravity and mass

Mass is affected by gravity. We measure the pull of gravity on mass and call it weight. Mass and weight are not the same thing, though the words are often mistakenly used interchangeably due to the consistent effect of gravity on earth. At this stage we won't talk of 'heavier' or 'lighter' rather that it has 'more mass' or 'less mass'.

We begin by making a simple scale to show that the more mass an object has, the greater the effect of gravity.

Next, we compare two objects of the same size but different mass. The objects' masses are different due to their differing densities. We do this experiment as a prelude to looking at the concept of 'density'. You may need to refer back to this experiment in the next section to be sure the concept is clear.

Following that, we move on to conventional weighing devices which operate by principles that are now understood by the student. At this stage we can introduce the concept of weight, the measure of the pull of gravity on an object. The greater the measured weight, the greater the mass must be. If we were to use the same equipment on the moon, where the gravity is less, we would find the same objects weighed less even though their mass remained the same.

Fill containers of the same size which you used for your sample Roman road, up to the same volume as the road, but with different, uniform objects such as rice, dried beans, water, coins, etc. Weigh each one and fill in the chart. Compare the different weights per the same volume. You may find that your road sample is heaviest. This is leading up to the final lesson in this section, which is density.

teacher's notes (cont'd)

density

In this experiment, we get the idea of different densities, and the idea of density being mass packed into a space. The aluminium foil can be packed more densely than the paper. This is because of the properties of the aluminium being more malleable, or bendable, and the paper being more stiff. In the extension, when you wet the paper, you'll find you can pack it more tightly because the quality of the paper is being broken down.

The Roman roads were constructed of very dense material, which was optimum for their function. You can calculate the density of your sample Roman road as described by the formula: **density = weight ÷ volume**. Both these values have been calculated in the preceding experiments.

Project 4 **Roman arch**

forces

This experiment introduces changing and moving mass. The forces experiment demonstrates three things:

1. left on its own the mass will not move; (apart from some very subtle flattening of the dough).

2. to push the mass along the table, one feels the friction between the table and the dough slowing the movement (the rubbing force); and

3. when it goes over the edge, gravity (the falling force) takes it to the floor.

In the Roman arch you have made, you sanded the blocks to make them more rough so that frictional forces would help hold the blocks in place. Friction operates on contact. Gravity helps to hold up the arch because when you tip the arch up to vertical, the blocks are pulled down by gravity. They "fall" into a closer fit i.e. they fall into a tighter semicircle and are compressed. Gravity acts over a distance, and also causes a compression force to hold the arch in place.

Note that the invention of concrete was vital to the building of arches. It is sturdy yet lightweight (i.e. low density).

friction

Friction (the rubbing force) is best initially demonstrated by rubbing your hands together. You can feel the heat generated by friction. This experiment shows the effect friction has on an object. In making the blocks for the Roman arch, friction was used to rough up the surfaces and also to change the shape of two blocks.

Friction is apparent in the Roman arch, by the blocks holding together, and between the ground and the "feet" blocks. Friction is helpful in these ways, and harmful in that over time, the blocks of many Roman buildings have been worn away and they have collapsed.

Friction operates through contact. To increase the effect of friction, you could have blocks with larger surface areas.

Project 5 **pile driver**

The Romans adapted the pile driver to use in the construction of bridges. The log, or pile, was set on the front and then pounded into the ground by dropping a heavy weight on it. The pulley at the top enabled the heavy weight to be lifted. Then the full force of the falling weight drove the pile into the ground. This construction device may have been used in warfare, when armies needed to quickly construct a bridge for strategic reasons, or for moving their armies from place to place.

gravity 1

Gravity (the falling force) is actually the attraction between two masses: the object and the planet, where due to the vast differences in the amount of mass each has, the object appears to 'fall' to the earth.

In gravity 1 we observe that the longer an object falls the more its velocity increases. This is evidenced by the increasingly dramatic effect on the object when it collides with the planet. It has more kinetic energy, therefore more energy is dissipated as the

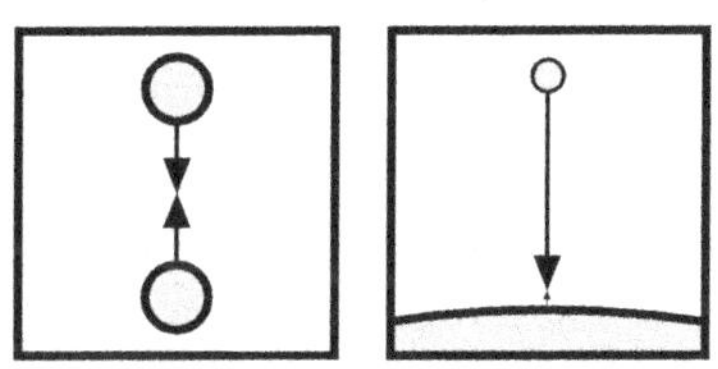

teacher's notes (cont'd)

motion stops. This is demonstrated by the distortion of the shape of the dough.

In the pile driver, when the mass is lifted and then you let go, the force of gravity causes it to fall. Because of its large mass, it has a lot of kinetic energy and when the mass hits the pole, the kinetic energy is transferred to it and it is driven into the ground. Frictional forces then hold it in place.

pulleys 1

Set this one up on some swings or parallel bars. In this exercise you may hear the remark, "It's easier to lift the bucket with my arm than with the pulley". This may be the case. However, the pulley may allow someone to lift something to a height which would otherwise be impossible for them to achieve. It is important to ensure that other students are well away from the lifting area, as a bucket of water dropped from a meter or two is dangerous.

pulleys 2

Depending on the type of pulleys you are using, you may need to use some extra rope or wire to keep the moveable pulley in the correct position. Be sure to practice this one first.

Vitruvius would probably *not* have used double pulleys because the full force of the falling mass was needed to drive the poles into the ground. The double pulley system would have absorbed some of the kinetic energy from the fall.

pulleys 3

It is very important to keep your technique accurate in this experiment. You should get very clear results which show that for the double pulley, the rope is pulled twice as far. This is where the ease of lifting comes from. The rope is pulled with half the effort, but twice the distance.

To advise Vitruvius, look at the big picture. Look at all of Vitruvius' resources. For our friend Vitruvius, a single pulley system would be quickest to set up and simplest to operate, and those may have been important considerations in his situation. The downside of a single pulley – that it takes more effort to pull it, especially for a heavy object, than does a double or multiple pulley – is offset by the fact that a team of slaves or animals could be used to do the work, and he was likely to have this at his disposal. If it happened that he had few men, slaves, and animals with him, you could consider recommending a double pulley system. (For a description of Vitruvius, see back cover.)

Project 6 **battering ram**

A battering ram was used by the Romans in wartime and also to demolish some of their buildings for reconstruction in Rome. The significance of the ram's head was important in that the ram is a persistent animal and when it decided to, it would butt up against something until it knocked it down. This was the result the Romans wanted and their "mascot" of sorts. The battering ram would be rolled into position and then with the strength of many men, repeatedly used to hit a wall with great force, using a swinging motion.

energy

Do this simultaneously – it is more fun and more dramatic! Everyone should hold the handles of their bag in one hand and twist ten full turns with the other hand. The potential energy of the twisted bag increases with greater mass put into the bag. This is observed as the greater kinetic energy of the spinning bag. The relationship of **more mass –> more energy** is the point we wish to make.

The forms of energy available for the early Romans to carry out work came from slaves, animals and inventions that used gravity. Inventions using wind and water power did not come on the scene until late in the Empire, as the slave populations decreased and the Roman population increased. It was then more economical to develop wind and water powered machines.

teacher's notes (cont'd)

moving and storing energy 1

PENDULUM

As you hold the pendulum up before letting go (potential energy), it has no movement. Once you let go, it travels and the potential energy is converted into kinetic energy. It is fastest at the bottom of the swing, closest to the table (kinetic energy). At the top of the swing at each end, just as it switches directions to come back again, there is an instant of no movement, just potential energy.

For Vitruvius to use his battering ram most effectively, the wall should be as close as possible to the part of the swing where the kinetic energy is greatest. This is where the rope hangs vertically, and at the bottom of the swing. Observe the arc of the swing of your model battering ram. Even though it seems that the ram moves horizontally, you can observe that the path is a small arc. So you would position your battering ram up close against the wall, so the wall is hit with the greatest force, and the motion of the pendulum (battering ram) follows through up and into and beyond the wall. This knocks it down.

The length of the string or rope of a pendulum is what determines the period, or speed of the swing. It also relates to how much energy the pendulum has. The greater the length of the rope, the greater is the potential force which can be exerted. So having a battering ram on a strong, high scaffolding ensured the biggest force was used by 1) having a long rope from which to swing the ram, and 2) having a heavy mass to heave at the wall.

gravity 2

THE TOWER

The important thing here is to raise the book slowly. If you raise it quickly, you add momentum and the tower will tip before we are able to observe the effect we are seeking. Fill in the table and then show the students how to take an average if they don't already know. Your results should show a decrease in the height necessary to topple the tower as the tower height increases (this is known as an inverse relationship). The reason for this is that the center of mass of the tower is raised as the tower gets taller and it is then less stable and easier to topple.

Our bodies are very unstable with their height to width ratio. Our bodies do not fall over because our intricate feedback system (brain, nervous system, sense of balance) is constantly giving information to our muscles to readjust themselves to keep us on our feet.

An understanding of the concept of center of mass would have been important to Vitruvius.

If a tower was very tall, it could be more easily toppled. On the other hand, he would need to prepare for the rubble to fall and make preparations to get out of the way!

Project 7 **onager**

One weapon which the Romans used in battle was the catapult. There are several different types of catapult, one of which is called an onager. They are powered by torsion, which is produced by repeatedly twisting a rope. An arm which acts as a lever, is embedded between the layers of twisted rope. The arm is pulled back and when let go, a rock or something similar flies into the air and presumably hits the enemy target. At the end of the arm of an onager, there is a sling, sometimes made of leather and held to the arm by rope or sinew. Another type of catapult is called a mangonel, which instead of a sling on the end of the arm, has a bowl-shaped bucket or container.

The name 'onager' means mule, because when released, the machine gives a big kick.

PEN

Before you click the tab, the pen has potential (or stored) energy. Once you click the tab, there is kinetic (or moving) energy. This is transferred to the paper ball on contact, which then flies off.

When the arm of the onager is pulled back it has potential energy. When it is released, it has kinetic energy.

teacher's notes (cont'd)

RUBBER BAND LAUNCHER

For this experiment, it is very important to practice your technique, otherwise you will have rubber bands flying so inconsistently that you will have trouble reaching a conclusion. Also, it will take practice for both people to let go of their rubber bands simultaneously. Be persistent. You will be able to do it with practice. When you do, the result is simple and clear. Both rubber bands hit the ground at the same time. This is because the force of gravity (attracting the rubber band to the earth and the earth to the rubber band) is the same on both, regardless of a horizontal velocity given to the 'shot' rubber band, as long as they are dropped from the same height. This is why it is important to brace against chairs of equal height.

Three differences between the onager and rubber band launcher:

1. It is the elastic forces in the rubber band which propels it forward. The onager uses torsion, or twisting, of a rope (or, in your model, string) in order to propel a solid object forward.
2. The onager uses a lever to launch an object. In the rubber band launcher, it is the band itself which is launched.
3. The onager can be adjusted to make the rope more or less twisted, and therefore produce more or less force. The rubber band launcher cannot be adjusted.

Three similarities:

1. Both machines can shoot a light object approximately 4–6 m (4–6 yards).
2. Both machines operate by storing and releasing energy.
3. Both should be handled with care around other people, so no one gets injured!

levers

levers 1	tongs – class 3	garlic press – class 2
	scales – class 1	scissors – class 1
	stapler – class 3	nutcracker – class 2

levers 2 Have stations set up in separate areas and run it as a race between two teams to see which team can carry out all the instructions first.

The onager is a class 3 lever: the fulcrum is on one end, the load is on the other end, and the effort is in the middle.

Project 8 **rolling tower**

The rolling tower was used by the Romans in battle. It afforded protection from enemy fire and had the advantage of height. There were heavy doors at various levels which could be opened a fixed amount so the soldiers could attack from a vantage point and still remain protected. The tower was rolled into position by using logs for wheels.

wheels

Build a cart. Get some wood offcuts free from hardware stores, which are generally glad to give these to students and teachers for projects. Practice using a hammer and nails. Small hammers are a wise idea. Helpful hint: Use the nail to make the hole in the lid, then place heavy tape over it to cover the burr. Poke the hole again through the tape.

Check whether the hole is too small for the wheel to rotate freely by inserting the nail. If it is too tight, rotate the lid on an angle so the hole gets bigger. Be sure not to pound the nail in too far.

IMPORTANT NOTE TO TEACHERS

It is important to review the main points of the previous lesson at the beginning of the lesson. We are constructing a platform of experience, from which the students will be able to deduce the concepts. Repetition helps facilitate this process.

Always practice the experiments and demonstrations before you present them. Then be prepared, as often happens, for something unexpected to occur, or a question to be asked for which you are unprepared. GO WITH IT! Even if it means spending extra time on the subject, or going and doing some research and reporting back. Good science comes from us, not the equipment. If you are genuinely miffed by something, share it with the children and continue to experiment and investigate why. Then they will be part of the real journey of science.

Start saving these things up in advance:

equipment list

CONTAINERS

- identical small containers
- large rectangular take-away container
- metal lids
- lots of empty drink cans, cartons, milk cartons, empty toilet rolls

HOUSEHOLD ITEMS

- measuring cup
- tongs
- garlic press
- scales
- plastic cup
- kitchen scale
- coins
- aluminium foil
- ice cube tray
- marbles
- sewing bobbin
- string
- rope
- buckets
- plastic zip-lock bag
- string, cotton
- shears or scissors, garden clippers
- clothespin (with spring coil)
- nut cracker
- egg cartons
- BBQ skewer
- waxed paper or baking paper
- rolling pin
- plastic drinking straw
- shoe box
- chairs, 2 of same size
- book, large hardcover

HARDWARE

- twine
- rubber mallet
- flat head nail
- tape measure
- needlenose pliers, small
- hammer
- wooden board
- masonite placemat
- washers
- meter-stick
- short nails
- pulleys
- gaffer tape, duct tape
- nuts & bolts
- sand paper (rough and smooth)
- dowel

FOOD

- rice
- sugar
- potato
- hard-boiled egg
- dough
- onion
- apple
- orange
- nuts in shell (e.g. walnuts)

TOYS

- a toy animal head with a flat back
- plasticine
- golf ball
- wooden blocks

STATIONERY / CRAFT

- binder clips, 15mm (3/4″)
- craft matchsticks (headless)
- 15 cm (6″) ruler
- 30 cm (12″) ruler
- ballpoint pen (click pen)
- pencils, circular in cross-section
- pencils, hexagonal in cross-section
- paper clip
- hole puncher
- cardboard
- chalk
- crayon
- poster board
- soft pencils
- pen
- lots of popsicle sticks
- bees wax
- paper
- plaster
- scissors
- recycled paper
- marking pen
- rubber bands
- sticky tape
- craft glue
- clipboard
- masking tape
- stapler
- whiteboard or blackboard, small

ROCKS

- large pebbles
- coarse gravel or marble chips
- sand
- smooth decorative stones
- rocks, e.g. quartz, pumice, etc

conclusion

Now that you have some understanding of Roman Machines and the physics principles on which they are based, you can if you wish, begin to share your adventure with parents, family, or others at your school.

Set up a large table where you can display and demonstrate your models. Put file cards with short explanations beside each model. Then, you could even construct a road by laying several model sections of road end to end along the table. Bring in a collection of figurines and set up two opposing armies (the Romans and the Goths). Have each side display their flag from their rolling tower. Try to shoot paper balls at the tower and land them inside the tower (remembering there should be no one standing behind or near that tower). Keep score of your successful attacks with your abacus and record the results on your wax tablet. Construct an arch, and try to knock down the enemy's arch with your battering ram. And use the pile driver to sink some straws into the sand to use for some strategic military purpose.

Based on what you have learned, let your imagination free to create, invent and improve your models just as the Romans did. We hope you have enjoyed this journey through the Roman Empire as much as we have enjoyed bringing to you.

For more *Primary Physics* books, visit us on the web at: www.primaryphysics.com.

About the author

Marti Ellen has a Master of Science degree from Stanford University, USA, a Diploma of Education from the Sydney Institute of Education, Australia, and is the author of several published scientific research papers. She has taught physics to young children since 1985. Marti has researched and developed the Primary Physics concept, and since 2000 has trained students, teachers, and academics, and promotes the teaching of physics to young children in Australia and around the world.

About the illustrator

Andrew Davies is a graphic designer and illustrator. He has a Bachelor of Design in Visual Communications from the University of Technology, Sydney and is the author and illustrator of *The Big Brain Book, Where's Zayn* and *Simply Drawing*.

About Sunshine Educational Pty Ltd

Sunshine Educational was established in 1996 and incorporated in 2011. The company is dedicated to helping teachers around the world teach physics in a fun and interactive way, thereby creatively preparing children for challenges in their future.

About the research and development of ancient machines

The models reproduced in this book are based on the machines created by skilled craftsmen in the Florentine workshop of Niccolai Teknoart SNC (Firenze), headed up by father and son team Carlo and Gabriele Niccolai.

For many years, their strong passion for history and engineering together with the techniques they have developed in their unique laboratory, has led to the creation of approximately 500 machines – meticulously reproduced, using materials of the era such as wood, iron, cloth, ropes and bronze. These reconstructed ancient machines have been on display in museums all around the world and are featured in encyclopaedias, scientific magazines and documentaries.

Precision in the reconstruction of ancient Roman machines has been achieved by the combination of intense research, careful studies of the ancient writings found in *De Architectura* by Roman architect and engineer Vitruvius, and the advancement in computer and software technology – which has enabled the testing of these ancient machines, reproducing them to scale and, as *fully working* machines.

Made in the USA
Monee, IL
12 January 2022

88332417R00037